Integration of Mechanical and Manufacturing Engineering with IoT

Scrivener Publishing
100 Cummings Center, Suite 541J
Beverly, MA 01915-6106

Publishers at Scrivener
Martin Scrivener (martin@scrivenerpublishing.com)
Phillip Carmical (pcarmical@scrivenerpublishing.com)

Integration of Mechanical and Manufacturing Engineering with IoT

A Digital Transformation

Edited by

R. Rajasekar
C. Moganapriya
P. Sathish Kumar
and
M. Harikrishna Kumar

Scrivener
Publishing

Wiley Global Headquarters
111 River Street, Hoboken, NJ 07030, USA

For details of our global editorial offices, customer services, and more information about Wiley products visit us at www.wiley.com.

Limit of Liability/Disclaimer of Warranty
While the publisher and authors have used their best efforts in preparing this work, they make no representations or warranties with respect to the accuracy or completeness of the contents of this work and specifically disclaim all warranties, including without limitation any implied warranties of merchantability or fitness for a particular purpose. No warranty may be created or extended by sales representatives, written sales materials, or promotional statements for this work. The fact that an organization, website, or product is referred to in this work as a citation and/or potential source of further information does not mean that the publisher and authors endorse the information or services the organization, website, or product may provide or recommendations it may make. This work is sold with the understanding that the publisher is not engaged in rendering professional services. The advice and strategies contained herein may not be suitable for your situation. You should consult with a specialist where appropriate. Neither the publisher nor authors shall be liable for any loss of profit or any other commercial damages, including but not limited to special, incidental, consequential, or other damages. Further, readers should be aware that websites listed in this work may have changed or disappeared between when this work was written and when it is read.

Library of Congress Cataloging-in-Publication Data

ISBN 978-1-119-86500-1

Cover image: Pixabay.Com
Cover design by Russell Richardson

Set in size of 11pt and Minion Pro by Manila Typesetting Company, Makati, Philippines

Printed in the USA

10 9 8 7 6 5 4 3 2 1

This book is dedicated to the profound memory of our beloved friend
Mr. R. Manivannan

31st October 1989 - 4th July 2022

Contents

6 Opportunities: Machine Learning for Industrial IoT Applications 159

Poongodi C., Sayeekumar M., Meenakshi C. and Hari Prasath K.

Preface

The Internet of Things (IoT) changes the way in which products are designed, prototyped and manufactured. In recent times, the core mechanical companies have experienced a strong transition on controlling mechanical systems by software-driven tools. Technology driven platform transform products into IoT-led smart devices, which can communicate with the producer, when they departed from the manufacturing line. Manufacturers should upgrade their production strategy by linking with IoT in order to maintain a long term sustainability.

This book is intended to compile and broadly explore the latest developments of IoT and its integration towards mechanical and manufacturing engineering. The book is envisioned to explore the fundamental concepts and recent developments in IoT & Industry 4.0 with special emphasis to the mechanical engineering platform to such issues as product development and manufacturing, environmental monitoring, automotive applications, energy management and renewable energy sectors. Topics and related concepts are portrayed in a comprehensive manner so that readers can develop expertise and knowledge in the field of IoT. It will provide professionals and students with a resource on the basic principles and application of IoT in manufacturing sectors.

We thank all the authors for their valuable research inputs. We render our sincere thanks to Scrivener – Wiley publishing team for their help with this book.

R. Rajasekar
C. Moganapriya
P. Sathish Kumar
M. Harikrishna Kumar

Evolution of Internet of Things (IoT): Past, Present and Future for Manufacturing Systems

Vaishnavi Vadivelu[1], Moganapriya Chinnasamy[2], Manivannan Rajendran[3], Hari Chandrasekaran[1] and Rajasekar Rathanasamy[3]*

[1]Department of Management Studies, Kongu Engineering College, Erode, Tamil Nadu, India
[2]Department of Mining Engineering, Indian Institute of Technology Kharagpur, West Bengal, India
[3]Department of Mechanical Engineering, Kongu Engineering College, Erode, Tamil Nadu, India

Abstract

The Internet of Things (IoT) is a platform that permits communication between gadgets, elements, and other digitized resources that send and receive data automatically without involvement of personal interactions. The key feature of IoT is the massive amount of information generated by finished systems that must be interpreted in the cloud in a short amount of time. The current study reveals the origin of IoT, brief revolution of IoT over two decades, emerging technologies with IoT, its current applications, and its future challenges. The current part of chapter starts with a broad review of the IoT revolution. After that, the technical aspects of IoT enablement technologies, protocols, and applications. The current chapter goal is to provide a more comprehensive overview of the most important protocols and application issues, allowing researchers and application developers to quickly grasp how various protocols interact to deliver desired functionalities without having to wade through RFCs and standard specifications.

Keywords: IoT, EC, manufacturing system, cloud computing, IoT architecture

*Corresponding author: rajasekar.cr@gmail.com

R. Rajasekar, C. Moganapriya, P. Sathish Kumar and M. Harikrishna Kumar (eds.) Integration of Mechanical and Manufacturing Engineering with IoT: A Digital Transformation, (1–40) © 2023 Scrivener Publishing LLC

1.1   Introduction

The current world is being pushed by the Internet and digital era, which is affecting digital life notions. The online and wi-fi networks have performed a crucial part in the digital era. In today's fact changing world, the major goal is to use modern technologies to minimize human-machine contact efforts while increasing machine-machine interaction capabilities [1]. Information and communication techniques has formed a trend in this regard, bringing concepts such as wireless control, remote monitoring, and so on, reducing the strain on humans and workers. An innovative and revolutionary technology known as the IoT was launched as a result of developments in wireless communications, cognitive computing and network connection [2]. Wireless sensor networks (WSN), bluetooth, long-term evolution (LTE), radiofrequency identification (RFID), near-field communication (NFC), as well as various wireless modern communications link things to the Internet. As a result, IoT might be described as "things connected via the Internet" [3]. The IoT links trillions of electronic things and electronic objects to develop a digital environment that allows humans to use new cyber technology to sense, analyse, regulate, and improve old physical manual systems [4]. A shipyard's operating efficiency was improved using Industrial Internet of Things (I-IoT) principles [5, 6].

Over the last 20 years, it has attracted a wide spectrum of audiences from business and academics, expanding the technology to a variety of scientific applications in various industries. IoT principles are applied in various fields like agricultural, supply of water, smart grid and energy savings, handling equipment and materials, industrial businesses, and transport planning. The Internet of Things concept has gained widespread acceptance and use in a variety of sectors. However, the IoT-related study findings do not delve into the IoT's fundamental development principles and research trends. Few studies had been done to disclose the beginnings of the IoT, analyze its popular study themes with the emphasize the challenges that the IoT will confront in the future. Furthermore, the advancement of IoT technology is inextricably linked to the support of associated theories and methodologies, and a rising number of researchers and practitioners are keen to learn more about the IoT's current state of development through reading publications.

1.2   IoT Revolution

The word Internet of Things was changed for M2M (Machine to Machine), which was anticipated and referenced by MIT professors who characterized the future world of communication in the late 1990s. Briton Kevin Ashton, a creative developer, was making a presentation for Procter & Gamble in which he presented IoT as a system that connects multiple devices using RFID tags for managing their supply chain. The adoption of RFID has enabled for the direct flow of information between devices to be accelerated. He envisioned a vision of data being collected, analyzed, and transferred with the absence of human interference. The IoT is a network system enabled by the Internet that aims to create real-time interaction between objects, machinery, and people using a number of advanced techniques. Likewise, a number of significant advancements aided in the IoT's development are depicted in Figure 1.1.

The first was an Internet-connected refrigerator developed by LG Electronics in 2000, which allowed customers to purchase online and conduct video chats. Another significant advancement was the creation in 2005 of Nabaztag, a little rabbit-shaped robot capable of providing up-to-date news, weather forecasts, and stock market updates. The Auto-ID center sponsored 103 branches throughout the world and created a standard to maintain the smart package to communicate with the other networks at distributors and buyers. Over time, the market improved, investments improved, chips improved, and chips grew cheaper and cheaper.Nest Labs was the first company to develop a sensor technology based, wireless based,self education, thermostat and smoke alarm to introduce IoT in 2010. The IoT was ultimately brought to the public's attention when, Google bought Nest Labs in 2014 and debuted the Amazon Alexa and, later, Google Home. Since then, the sector has been growing faster.

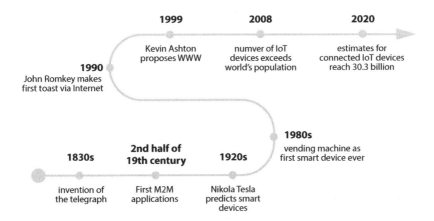

Figure 1.1 Power generating capacity installed in 2017 [5, 6]. https://www.avsystem.com/blog/what-is-internet-of-things-explanation/.

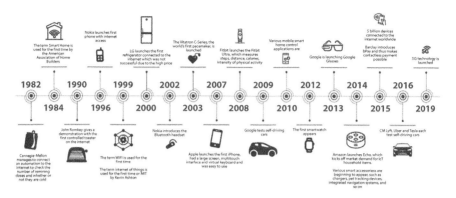

Figure 1.2 Power generating capacity installed in 2017 [5, 6].

IoT has gotten a lot of attention during the last 20 years, with a lot of government officials, business leaders, and academics believing that this essential technology is evolving people's standard of living and surrounding conditions are depicted in Figure 1.2 [7–10]. Several researchers have stretched IoT associated research with service as Internet of Service [11], equipment as Internet of Machine [12], humans as Internet of People [13] and information as Internet of Knowledge [14, 15]. With the advancement of science and technology, IoT is expected to have a wide range of applications in the government service and domestic sectors [16]. The application of IoT helps to reduce the pollution created by human activities and enhance the economic growth of the country [17–19]. To apprehend this possible development in the economic, the development of diverse developing technologies and services must keep up with the expansion of market demand [14, 20, 21]. The IoT diversified developments in the aspect of technology, applications, undamental ideologies, design aspects, and trends in the anticipated growth are combined with interdisciplinary association with the telecommunications, electronics and informatics [7, 9, 22]. After the two decades of development in IoT, the research have been extended, accepted in numerous industries and employed in smart medical care, smart agriculture, smart supply chains and smart cities [23–28].

1.3   IoT

The advancement of mobile gadgets, automobiles and integrated device has aided in the creation of a smart world of linking the gadgets that can perceive, gather data, cooperate, and make choices without the need for human intervention [29]. This intelligence is called Internet for things, which simplifies the day-to-day life of humans. The IoT is characterized as a dynamic, self-configuring manual linkage and virtual devices connected through interoperable communication, media and standards [26, 30, 31]. Wi-Fi, Bluetooth, Zig Bee and other protocols are used by these items to communicate with one another. The interconnection of numerous communication technological innovations which includes wireless network and sensors, controller networks, tracking and identification networks and so on to promote interactivity and cooperation between them is a crucial component of the IoT [32]. Some emerged real-time examples are wearable fitness and trackers (like Fitbits) and IoT healthcare apps, voice assistants (Siri and Alexa), smart automobiles (Tesla) and smart appliances.

1.4   Fundamental Technologies

Hardware such as sensing devices, electronic controls and integrated sensor hardware; software components such as data storage requirement and information processing analysis of data predictive analysis and visualization; and demonstration as unique, simple to use visualization and explanation tools technique that can be accessible across multiple platforms and built for a variety of implementations fields are the three IoT components that allow for smooth widespread computing [9]. The current part discusses a few enabling technologies that contribute to the components of IoT.

1.4.1   RFID and NFC

The key innovation in RFID technology is the design and development of a wireless microchip for the embedded communication paradigm for data transfer [33]. It provides the automated identification of whatever is connected to this electronic barcode [34, 35]. RFID devices, often known as RFID tags, which used microchips to transmit data wirelessly. RFID tags emit data over the air, and an RFID reader recovers the signal, allowing items to be identified based on the data received (barcode). The most often utilized device for IoT applications in many industries such as retail, supply chain, healthcare, banking, privacy control, and social applications [36]. NFC is a quick sequence of high frequency network technology that exchanges data between a few centimetres, making it easier for people to use their phones and providing a variety of loyalty applications such as locking and unlocking doors and cars, exchanging contact information, paying for general populace transportation, reading newspapers, and others [37].

1.4.2 WSN

WSN is a broad system of intelligent detecting device that gather, process, analyse, and transmit information [38]. The WSN is made up of the components listed below. To begin, the gather unit is referred to as a sensor, and it is responsible for collecting data in the form of waves and converting it into electronic knowledge that the formulating unit can understand. The second component is the processing unit, which is in charge of analyzing the recorded data. The transmission unit, which is in charge of all data transmission and receiving, is the third component. Finally, there is a power main controller that is a critical network component [39]. WSNs are also employed in a range of applications, including monitoring systems (e.g., surveilling pollution, disasters, and wildfires), industrial (e.g., intelligent lighting control, automation), defence, and healthcare applications [40].

1.4.3 Data Storage and Analytics (DSA)

A big storage unit is necessary since a great volume of data is produced and exchanged in IoT. As a result, data storage has become an important issue in the Internet of Things. To ensure effective communication, digital cities, intelligent and interconnected societies, and better medical are just a few of the data processing and storing technologies available. Cloud-based information management and processing have grown in popularity in recent years, and they are widely used and desired because they can speed up information analysis and deliver a highly secure interchange of information [38].

1.5 IoT Architecture

To manage the trillions and millions of linked devices, the IoT requires a flexible layered architecture. The diverse models originate from an increasing diversity of latent designs [41], and IoT-A is attempting to establish a common architecture to meet industrial demands, which is being investigated by the researcher [22]. Three-layer designs are the most typical IoT architectures because they are versatile, practical, and simple execution. The perception layer, network layer, and application layer are the three tiers of the architecture. Figure 1.3 depicts a three-layer IoT design, with an explanation underneath.

- The perception layer has sensing capabilities, which means it gathers as well as accumulates specific data about the surrounding environment in which digital things are present.

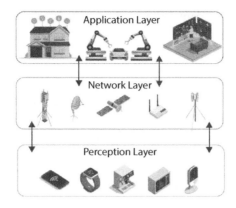

Figure 1.3 IoT Architecture from [32].

- The network layer is accountable for qualifying transmission of data and processing across several gadgets connected through the Internet.
- The major function of the application layer is to deliver the ultimate user with a particular service relying on the specific application.

1.6 Cloud Computing (CC) and IoT

The growing emergence of interlinked devices throughout the world and changes in users and applications demand for massive amounts of data processing, prompted the development of a set of unique technologies that enable quick data processing and dependable services. One of these technologies is CC [32]. CC is a highly adaptable and extensible technology that enables a variety of services for IoT applications. Cloud is an innovative platform which aims to deliver a various service to the final customers [42]. The services include such as data storage choices, software strategies and analysis, a suitable platform, and fundamental development infrastructural facilities. Private, public, hybrid, and community clouds are all examples of cloud implementations. A large number of people use the public cloud via the Internet. The private businesses utilize the private cloud and designed a community cloud for specific set of organizations [43, 44]. And hybrid cloud is designed to control the cost and issues related to control [45]. The cloud capacity to manage enormous volumes of information utilizing data analytics, visualize the huge data by users and machine learning [46]. Specific required individual service, comprehensive network access, resource sharing, fast adaptability, and quantifiable services are some of the unique qualities of CC [47].

- Self-service on requirement without needing human interaction, a customer can provide computer competences includes as network storage and processing time are desirable.
- Broad network access refers to the availability of a wide range of network capabilities that may be accessible through standard processes to encourage the utilization of diversified thick client platforms such as mobile, laptops, tablets and workstations.
- Storage, compute, memory, and network bandwidth are pooled to serve several users in a multitenant paradigm in which physical and virtual resources of various types are dynamically assigned and reassigned based on demand.
- Rapid elasticity refers to the capacity to dynamically allocate and release resources in response to demand at any point in time.
- To automatically control and optimize resource utilization, cloud systems use metering capabilities at some level of abstraction, such as storage, processing, bandwidth, and active user accounts.

1.6.1   Service of CC

The cloud system utilized the specific type of service which is the primary categories of cloud architecture. The cloud architecture is described in Figures 1.4 and 1.5 with explained below;

- Platform as a Service (PaaS): A toolkit makes the PaaS available to application developers as a development environment. PaaS allows one (user) to deploy self-created or bought apps into cloud infrastructure utilizing the provider's tools and programming language.
- Software as a Service (SaaS): The SaaS paradigm simplifies the utilization of cloud-based application to access the variety of utilizer gadgets via a tinny buyer line such as a web browser or a programme interface.
- Infrastructure as a Service (IaaS): According to the Internet Engineering Task Force, an (IaaS) distributer's delivers

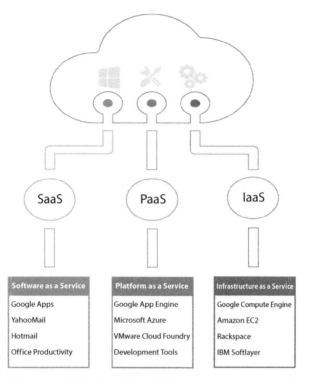

Figure 1.4 Architecture of CC adopted from [32].

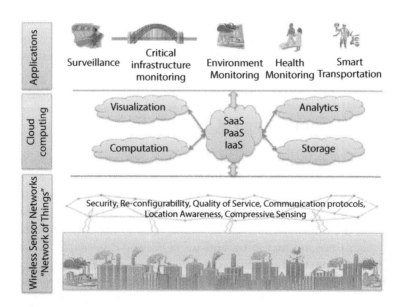

Figure 1.5 Architecture of CC adopted from [9].

online or physical systems/equipment and other resources. One can have access to processing electricity, storage facility and further computer assistance with IaaS.

1.6.2   Integration of IoT With CC

IoT and CC are dual major knowledge that donate to human routine lives by allowing IoT users to access a variety of services. Researchers are focusing on integrating the cloud and IoT, which have both realized quick growth and may be regarded complimentary to one another [48–51]. The incorporation of IoT and cloud would have high storage, processing and networking capabilities under a feature of IoT. The types of devices, technologies, and protocols of IoT utilize the scalability, interoperability, reliability, efficiency, and other IoT qualities are difficult to achieve. The integrating the IoT with cloud would overcome the above difficult concerns and have other benefits like ease-of-access, ease-of-use, and so on [52, 53]. The cloud could be a best part in handling the IoT services and creating the applications and services [48, 54].

The Cloud IoT paradigm, which combines cloud and IoT interoperability, aids in the development of new cloud-based intelligent applications and services, such as Sensing as a Service (SaaS), Sensing and Actuation as a Service (SAaaS), Sensor Event as a Service (SEaaS), Sensor as a Service (SenaaS), DataBase as a Service (DBaaS), Data as a Service (DaaS), Ethernet [55–57]. Healthcare, smarter house and digital monitoring, video surveillance, automobile and digital mobility, smart logistics, and environmental sensing are just a few of the applications that have evolved from these services [54, 56, 58–61].

1.7   Edge Computing (EC) and IoT

New applications, such as streaming video and virtual games, necessitate a quick response time [62, 63]. Similarly, energy usage is a major concern in wireless communication due to the limited resources of IoT devices, which are unable to perform calculations and computing locally. Moreover, the information must be transmitted to the cloud data centre, which takes time and may affect end-user QoS and experience. EC aspires to become the next IoT solution, capable of overcoming a variety of challenges such as duration-constrained and compiler optimization applications. EC denotes that enabling technologies that enable computing at the network's edge, on both downstream and upstream data for cloud and IoT applications [62]. Any computational and network resources positioned between data sources and cloud data centres are referred to as "edge." The edge between body things and the cloud, for example, is a smart phone; the edge between house things and the cloud is a gateway in a smart home; and the edge between a mobile device and the cloud is a mini data centre and a cloud let [64]. EC in the field of Industrial IoT adds agility, real-time processing, and autonomy to produce value for intelligent production [65]. Minimizing networking load and transmission delay, separating the monopoly of big inventors, delivering small and medium inventors with each and every possible chance to help cultivate advance innovations, minimizing energy consumption of mobile nodes and abolishing congestion within the core network as well as allowing greater durability, security and privacy protection are some of the benefits of processing data at the network edge [66, 67].

1.7.1 EC with IoT Architecture

EC refers to the activities of IoT devices at the network's edge or limit that are linked to the distant Cloud [68]. Due to the huge increased amount of IoT devices, numerous organizations offer several IoT designs from various angles, and EC was already highlighted as an essential guidance for IoT systems [9, 69, 70]. The latest research on the subject attempted to demonstrate that EC architectures are the best solution to decrease latency, improving secrecy, and cutting bandwidth costs in IoT-based applications. The EC with IoT architecture is developed by [71] and shown in Figure 1.6.

The edge-computing IoT architecture is made up of four primary components such as the IoT final device, cloud, the edge and users. The architecture is designed with both available resources and the individual characteristics of each party in mind. Users use sophisticated IoT apps to improve their lives simpler, and instead of interacting directly with IoT end hardware, they frequently contact with them via interactive interfaces provided by the cloud or the edge. End devices for IoT are firmly implanted in the real environment. The IoT end device monitor the lively environment and perform actions to regulate it, but they lack sophistication in computation-intensive activities. Although the cloud has essentially infinite resources, it is frequently located far distant from end devices. As a result, a cloud-centric IoT system is unlikely to perform well [72], especially where actual data is required. Because the edge is such an important element of the entire architecture, it may integrate the two and coordinate the other three parties to collaborate and complement the cloud and IoT end devices for the best possible performance.

Figure 1.6 EC with IoT architecture [71].

1.8 Applications of IoT

In terms of adaptability, the IoT offers enormous opportunities for societal, environmental and economic implications. IoT have been adopted in various potential fields such as social, environmental, and economic. The fields like mobility, smart grid, industrial processing, agriculture and breeding, smart residential, medical and healthcare, societal security and environment production, and independent living. The above-mentioned applications are utilized by everyone in their day-to-day life. The usage of the above-mentioned applications is very essential and gain the much importance in the recent years. Figure 1.7 demonstrates the applications of IoT in various different fields.

1.8.1 Smart Mobility

Smart Mobility is a method of enabling seamless, effective, and flexible mobility throughout several modes. Due to the passage of time and the rising need of society, Vehicular Ad-hoc Networks (VANETs) have grown and gained a great deal of recognition. As a result, one could argue that it marks a fundamental shift toward a highly adaptable and cross transportation

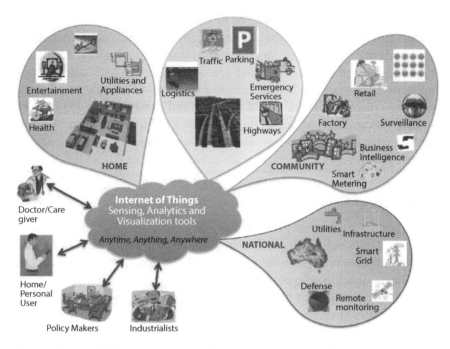

Figure 1.7 An IoT diagram depicting application from various sectors based on data from [9].

network. The Internet of Vehicles (IoV) is a new technology that promises to enhance road safety by safeguarding or minimizing incidents and allowing new optimum mobility modes. Many scholars have tackled additional challenges, such as traffic and travelling from one location to another, in an effort to improve [3].

The various developments from the researchers are IoT below smart mobility field with technical challenges associated to wireless communications [73]; development of Advanced Driver Assistant Systems (ADAS) to provide a direction and arranging and rectifying the multimedia and connectivity issues systems [74]; smart city and smart mobility is created to handle the Eco-Conscious Cruise Control for Public Transportation [75]; development and demonstration of virtual real objects, such as sensors are interrelated with the Virtual Object (VO) framework to monitor the transport [76]; electronic vehicle is connected with the electric vehicle collaboration into the Internet of Energy (IoE) for the Smart Grid infrastructure [77]; cloud with IoT [78]; cloud integration [79]; handled the whole network was affected by road traffic delays at one place [80, 81]; use of mobile sensors to identify a variety of metropolitan transportation modes might generate data that could then be categorized to better understanding the amount of behavioral alterations in travel [82]; applications for safety is developed to deliver messages and warnings regarding accidents, curve speeds, traffic infractions, and pre-crash detection [83]; convenience applications is created to help in personal routing and provide guidance in handling critical situation, out of network and power failure [84]; smart parking system uses a wireless connection or a publish subscribe communication paradigm to notify smart automobiles about availability parking spots in the location [32].

1.8.2   Smart Grid

A smart grid is an electricity distribution system that monitors and responds to regional demand fluctuations using online communications technologies [85]. It's also recognized as a digitized technology which allows both-way communication, allowing ultimate consumers to request power after conducting examinations using sensors and receiving power in return. The grid distributes energies in accordance with the estimated demand, which has been established [86–88]. This information is analyzed by Supervisory Control and Data Acquisition (SCADA), a centralized server that issues urgent instructions, answers to modification requests to improve the stability, and protects the electricity system [89, 90].

The other allied developments to support the smart grid includes as measuring the energy efficiency, smart infra-structure and a forceful ecosystem [91]; implementation of effective monitoring system [92, 93]; creation of Automatic Meter Reading (AMR) and Advanced Metering Infrastructure (AMI) to record the utilization electricity without manpower [94]; integrating power generators (such as solar panels, wind farms, and other alternate energy sources) with the main power system to provide centralized control of the power network [95, 96].

1.8.3   Smart Home System

A Smart Home is a residential area that, like any other home, is outfitted with heating, lighting, and other technological devices. They can be controlled remotely using a smartphone or computer, which is a significant difference [3]. The smart home aims to provide adaptive monitoring of different electronic gadgets linked to houses, such as televisions, refrigerators, cookers, air conditioning units and so on [97]. Home automation services are used to control air conditioners, air cleaners, and curtain activities; residential protection facilities are used to prevent gas leaks and identify potential criminal activities; and residential managerial services are used to intelligently control smart devices such as cookers and televisions [98]. IoT is integrated with RFID [99, 100], cloud [101, 102] and EC [103, 104] to create a smart home system effectively.

1.8.4   Public Safety and Environment Monitoring

Public safety and environmental monitoring are the technique of conducting evaluations on seasonal changes, threatened species preservation, monitoring the quality of water and a variety of other qualities that are directly or indirectly connected to our environment. Various applications are associated with numerous sensors and other monitoring equipment to observe timely changes in environmental parameters [3]. The utilization of IoT to monitor the sensing qualities that open up an unimaginable market of possibilities [105]; security procedure is generated for monitoring the environmental by utilizing mobile WN [106]; utilizing RFID for monitoring the environmental [107]; Scanning of the environment is done through IP, with WSN for associating with many sensors, resulting in a rapid adoption, longer lifetime, excellent quality, and cheap maintenance [108]; established an Environment Internet of Things (EIoT) to track important factors such as soil, air, water and wind [109]; environmental monitoring and management, a combination of CC with IoT, Geoinformatics (Geographical Information System (GIS), Global Positioning System (GPS), Remote Sensing (RS) and e-science are utilized [110].

1.8.5   Smart Healthcare Systems

Because of IoT-enabled devices, continuous remote monitoring in the medical business is now possible, allowing physicians to give better treatment while keeping patients safe and healthy. Patients' engagement and happiness have increased as interaction with doctors has become simpler and more efficient. IoT has altered the sanitary system and evolved it into a predicting and smart network by analyzing huge data and integrating various IoT devices to capture real-time physiological data of patients such as glucose levels in the blood, temperature monitors, and other critical data [111]. The IoT has the ability to relieve pressure on hygienic services while also offering tailored healthcare facilities to enhance people's standard of living [112].

Data management is a delicate topic in healthcare systems, since client information contains critical and confidential information that must be assessed and handled in a timely and safe manner [113]. Remote Patient

Monitoring (RPM) enables patients to be monitored irrespective of where they are and caregivers and family members may be contacted from a distance [114, 115]. Data acquisition, display and diagnostics are the three elements that make up the RPM system. Sensors are attached to the patients (for example, a blood sugar level) to gather data, which is then sent from the patient's android mobile device to the handling data unit (diagnostics component) for analysis. The IoT is a hot topic in current trends and also provide solutions to handle the difficulty of exposure to healthcare services, the growing ageing demographic with medical problems and their necessity for remote patient monitoring, rising hospital expenses, and a desire for telehealth in developing nations [112].

1.8.6   Smart Agriculture System

Smart agriculture system incorporates GPS-based remote monitoring, moisture and temperature sensing, intruder frightening, security, leaf wetness, and suitable watering facilities. It employs wireless sensor networks to continually monitor soil characteristics and environmental conditions [116]. Throughout the farm, many sensor nodes are put at various places. These parameters are controlled by through whatever separate device or broadband Internet service, and the activities are carried out by connecting sensors, Wi-Fi, and a camera with a microcontroller. Mostly wireless sensor networks are used for agricultures system for sensing the environmental parameters such as energy consumption, topology construction and cyber-attack security system [117]. Patil and Kale [118] have used sensor technologies and an IoT wireless network to understand the current state of an agricultural system. Remote Monitoring System (RMS) is offered as a combined strategy combining online and wireless connectivity. The primary goal is to collect actual data from the farming cultivation atmosphere in order to give quick accessibility to farming amenities which includes as cautions notice via Short Messaging Service (SMS) and guidance on climate conditions, crops, and so on.

Khoa *et al.*, [119] proposed a novel sensor node topology that emphasizes the use of low-cost, high-efficiency elements such level of water, moisture levels in soil, heat, wetness, and rainfall detectors. Furthermore, the transmission module used is based on LoRa LPWAN technology to assure effectiveness of the network. According to Naresh and Munaswamy [120], IoT modernization will aid in the collection of information on conditions such as climate, moisture, temperature, and soil productivity. Crop web-based examination enables the identification of wild plant, water level, bug location, creature disruption in the field, trim development, and horticulture.

Farmers are used by IOT to connect to his home from anywhere and at any time. Remote sensor structures are used to monitor homestead conditions, while smaller scale controllers are used to regulate and mechanize house designs. Kumar *et al.*, [121] have used an IoT-based application to monitor soil moisture utilizing a detection component, connecting through the online with Wi-Fi module, and controlling the switching of the submerged motor pump (motor driver-289D) using Arduino Uno R3.

1.9   Industry 4.0 Integrated With IoT Architecture for Incorporation of Designing and Enhanced Production Systems

Many industries have benefited from research on ICT-based production system during the last three decades, which has changed the conventional ways of designing, manufacturing, and distributing their goods [122, 123]. In industries, automated industrial operations are carried out using IoT-based systems and controls. It covers the entire process from the commencement of production through delivery of the finished product [6]. Figure 1.8 depicts the automated IoT manufacturing processes. Data from IoT-enabled manufacturing addresses connectivity, computation, and control challenges. Power stations, management of water resources, chemicals production, and materials fabrication all use IoT devices. The detectors are used to manage and regulate the production process. This could also collect actual information on the operation of industrial systems in real

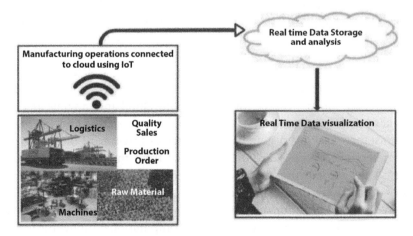

Figure 1.8 IoT-based manufacturing process from [6].

time. It assists businesses in enhancing quality and regulating industrial procedures. It also is specifically designed to support the manufacturing cycle more efficiently and speed up output.

Using group technology and Industry 4.0, manufacturers are increasingly focused on genuine tracing of manufactured products in an assembly line. AGent based Systems (AGS), mobile robots, data interoperability, Digital Twins (DT's) principles, machine learning techniques, and 3D printing are all used in Industry 4.0. Manufacturing companies face a variety of challenges when incorporating those certain technologies into their design and development phases, such as quick digital designing/redesigning of product lines, digitized recreating processes/products, production process planning and decision making, feature-based automatic machine formation, intellectual set - up/fixture planning, and supply chains/market delivery as required [123].

Kuo *et al.* [122] have created an algorithm that anticipate the working condition of spring manufacturing machines. They employed a multistage data collection and evaluation method to achieve Industry 4.0 workflow, in which the collected component values from processed data were anticipated using neural networks. Santos *et al.* [124] have developed a huge data structure model that was used at Bosch to accomplish Industry 4.0. The architecture of data gathering and analysis technologies using a big data analytics strategy, which includes web services, apps, databases, security and management. Similarly, Kumar *et al.*, [46] have utilized Simulated Annealing (SA) centred metaheuristics and Principal Component Analysis (PCA) to propose a big data solution to the facility layout problem. It is developed a dataset with 14 criteria based on the three V's (Volume, Variety, and Velocity) to quickly construct a defensible architecture that met their goals. Lin *et al.*, [125] have introduced a block-chain created reciprocal recognition system with perfect accessibility of management structure for Industry 4.0. This paper provides the BSeIn conceptual framework for implementing a flexible and changeable smart factory. The study established methods for evaluating data security across terminals, the block-chain system, the manufacturing network, the cloud and tangible assets.

Likewise, Dinardo *et al.*, [126] developed a proactive and intelligent circumstance tracking system based on regular tracking of the energizing parameters of frequency response acquired from the investigated equipment. A deeper learning method for detecting classical machining operations from CAD data was disclosed by Peddireddy *et al.*, [127]. They adopted modest grinding and rotating properties from Standard Triangle Language (STL) models to educate their computational model for feature detection. Kim *et al.*, [128] showed a depth restoration approach for pipe element catalogues employed in plant 3D CAD model reconstruction. In their work, they highlighted the reverse engineering concept of restructuring from 3D point cloud data/subdivided point cloud data, and they used depth learning algorithms to effectively generate the models required.

1.9.1   Five-Stage Process of IoT for Design and Manufacturing System

In a digital world, a fully networked ICT created designing and production method need a powerful AI. Furthermore, to accomplish and apply the numerous manufacturing needs, the connectivity necessitates clever fully computerized AI utilization of smart programmes that can acquire, reason and create choices on their own. For that, AI-based 5-stage design is proposed and illustrated in Figure 1.9 (taken from [129]), which represents the integration of the different technologies stated above to accomplish Industry 4.0 with IoT.

Stage 1: RFID CAD data
The proposed RFID created method in the multistage architecture addresses the difficulty of speedy development/redevelopment by merging design and manufacturing data. The issue is that it is the source of a massive quantities of data throughout the procedures, i.e., it can identify

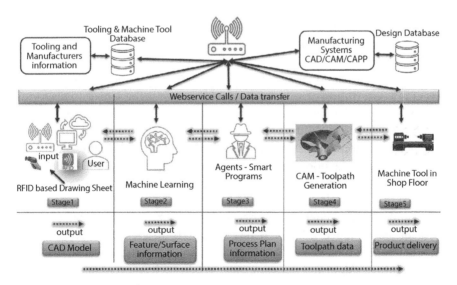

Figure 1.9 Interoperability of design and production systems in a multistage IoT-based industry 4.0 Infrastructure from the [129].

the process plan, machine status, tool paths, and availability, among other things, using product features/shapes and material (combinations) information. It will be performed in a digitally online platform and international digital CAD standards such as STEP will be required for RFID input. The concept will let us create an electronic simulator in which we will be able to apply IoT established information sharing and recovery technique to achieve Industry 4.0.

Stage 2: CAD model retrieval using web services and machine learning
The next step is to obtain and display details of data from websites server in the Graphical User Interface (GUI) after scanning the RFID smart technologies with the CAD model from beginning stage. The GUI communicates with a web application which utilize the Simple Object Access Mechanism (SOAP), which is a messaging protocol for sharing structured data for retrieving CAD models. During this procedure, the user interacts with the GUI and establishes a connection among the local network computer and a distant business/web server located at a given location. The RFID CAD relevant data is saved in their separate databases.

Step 3 & 4: Agent based Computer-Aided Process Planning (ACAPP)
In this study, sensors are utilized to overcome a range of complex difficulties that demand intellectual understanding and deliberate decision making, notably from stages 3 until 5 when procedure planning, toolpaths, and production schedules confirmations are necessary. Essentially, those sensors are small or ant colony size scripts that are used to solve a variety of complex problems that need advanced reasoning and better decision making. Essentially, individual sensor has four structures. The first, a belief set is a general relational model used to express agent beliefs. Second, an event is a description of an occurrence that requires the sensor to respond. Thirdly, a sensor's plan is identical to its activities. Those are the guidelines that the agent sensors must adhere to in order to achieve its objectives and manage the happenings that have been assigned to it. Finally, a competency facilitates the pooling and reusing of an agent's functional elements. A capacity is a collection of planned, happenings, belief sets, and other abilities that work altogether to give a sensor a specific talent.

Manufacturing generates a process plan for a basic identifiable component or shape called a "slot." It has the ability to generate feature-based manufacturing data as well as event for manufacturing characteristics, as well as cutting edges and machine data. When such an event or aim occurs, it determines what course of action to take and achieves the procedure effectively. The 'Plan' will then be carried out by retrieving the particular confidence from a database of tools and process parameters, exactly like a reasonable human would.

Stage 5: CNC machine compatibility with CAD/CAM software
In a production system setting, data transfer from CAD/CAM software to a CNC machine through a CPE is tricky [130, 131]. This is due to a variety of issues faced during machining, even by expert employees. Some of them arise during decision-making and tool path adjustment in situations like tool breakage during machining; incorrect toolpath creation from a CAM application software; scarcity of cutting tool and so on. The list will get longer since machining difficulty changes depending on other parameters such as CNC machine capabilities, component form, material, and so on. In many cases, a rapid change along with CNC programme or G&M codes is necessary, although this is complex and time-consuming. This is especially true when one of these events occurs often during practical manufacturing. There are CAD/CAM software along with CNC machines to assist, however the background of the issue shifts because if a new part with a non-machinable design needs to be machined in order to 'rapid manufacture' the product. The current architecture considers using IoT technological innovations to solve these challenges by obtaining real-time data and determining whether or not a digitalized twin of the entire process is achieved by utilizing system applications equivalent to a Raspberry Pi, transistors for I-IoT information exchange, cloud rail box, and so on.

1.9.2 IoT Architecture for Advanced Manufacturing Technologies

RFID is a wireless transmission technique that is implemented in a variety of applications, including shipping, tracking, and delivery. RFID is a contactless technology that detects things that have smart tags attached to them. Through communication with tag antennae, an RFID reader uses previous data from goods and the environment. A wireless device domain can be used by scanners and labels [132]. RFID can also quickly distinguish a variety of products. Furthermore, its anti-collision technology allows it to differentiate a specific number of tags at the same time. The RFID framework is highly robotized and offers a broad variety of adjustments for easy and universal computing applications such as position tracking, access management, and environmental monitoring [133]. An RFID digital tag, an RFID sensor, and a digital storage are three components of an RFID system, which equipment linked to RFID that is used to record data in the cloud. The antenna and chip of an RFID smart tag are wrapped with a transmitter, controller, and polymer-encapsulating substance. As soon as the RFID scanner finds a tag, it continues to search for RFID smart tags. The data from the tag was kept in the data storage to process later [132].

Several manufacturing organizations struggle to trace components from their suppliers in order to fulfil client orders during peak time of production and fail to deliver the product on time. The present plan collects and tracks data from numerous suppliers' factories in actual environments, such as procedure information, basic components, quality information and so on. Manufacturing business companies may manage their actual environment workspace manufacturing data includes production schedules, semi-finished goods storage, and input substances amount in process, as a result of the real-time statistics mining evaluation. The management of manufacturing organization now contains dynamic manufacturing records, which allows to analyze implemented manufacturing plan in real time, adapt quickly to changing market conditions and adjust production and purchase plans. The quality of production is handled dynamically and basic information regarding production quality and traceability can be provided.

Manufacturing organizations are started using RFID systems to provide solution to eliminate the manual coding at every level of the manufacturing process [134]. The use of an RFID built system decreases the possibility of scanner mistake in the manufacturing line. RFID tags are used by companies such as Vauxhall and BMW to customize client orders [135, 136]. A reprogrammable RFID card is configured and fastened to the production parts based on the client's requirements. The RFID card guarantees and tracks parts for right color, interior, and any customized extra features of a specific client during the manufacturing process [137].

1.9.3   Architecture Development

For produced items, the system's major purpose is tracking, wear estimation, quality of products, products categorization, and inventory monitoring. To identify premounted RFID tags on manufacturing items, RFIDs are connected to input module and output module in every production station, appropriately. The RFID reader recognizes the component as it passes into the location using the radio frequency identification mechanism. A vibration sensor is a component of the system that allows it to capture enormous amounts of vibrational data while drilling or milling. Smart RFID tag is a type of uniqueness that employs RF signal to detect and track smart tags [138]. The RFID based IoT architecture is adopted from [132].

(i) System Architecture

The system architecture for gathering and analyzing data during the production process are depicted in Figure 1.10. Information passes from the smart tag to controller and followed to rear-end cloud data base, allowing for real-time analysis and monitoring of the production process while maintaining the quality of the components. The system creates a map of the production procedures and updates unique data for each item, guaranteeing that the system is transparent and traceable.

(ii) Product Tracking System

By linking RFID and vibration detecting modules to the Internet through the MQTT protocol, manufacturing parts may be tracked in real time. As follows, the system analyses production information connected to the part and stores the information to assure the final quality.

- Time of arrival on the shop floor
- Process start up time.
- Duration of the process.
- Time of departure from shop floor.
- Constraints pertaining to shop floor

The manufacturing process begins with the manufacturer's request to the product provider. As soon as the request is

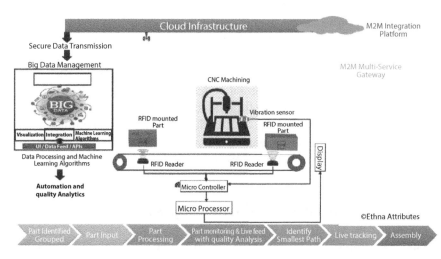

Figure 1.10 IoT architecture [132].

received, the supplier creates an RFID tag for the needed shipment and sends the part to a different production work floor. Once the component arrives at the work station, the RFID reader activates the vibrational sensor to use vibrational data to confirm the quality of the manufactured part. The RFID technology assists the controller in distinguishing each part and mapping data into the appropriate data base to improve part traceability. The system is capable of detecting the problematic part and rejecting it with the aid of a quality analysis system built in the same controller. To monitor and synchronize data, RFID tracking systems can be linked to cloud-based infrastructure.

(iii) Analysis of Vibration

The sensor configuration on the machine tool for recording vibration data. Tool wear assessments are used to develop and test signal processing algorithms. As a result, variations in vibration signatures may be recorded during milling operations over the tool life in order to extract a set of data that will be used as tool wear indicators.

(iv) GT-based categorization with RFID integration

Manufacturing parts are coded with numbers, letters, or a combination of the two in the classification technique, which is based on a classification system. This strategy is also known as the class and coding approach. Elements can be classified based on the following four characteristics, such as operations needed, forms, dimensions, and material tolerance requirement. Using the Optiz categorization and coding technique, each manufactured part is allocated a 10-digit code, with each digit representing a unique feature of the item. The code is classified into three major categories, such as form code, supplemental code, and secondary code. The form code contains the fundamental design features of each part, the supplemental code contains data about manufacturing and the secondary code contains more information about the manufacturing properties.

1.10   Current Issues and Challenges in IoT

Despite the aforementioned benefits and available solutions, different concerns and obstacles must be researched on leading edge of IoT applications.

Following, we explore the key issues, as well as new concepts and recommendations that the research community should take seriously.

1.10.1   Scalability

Scalability is an important aspect of every network architecture since the network should maintain a large amount of requirement, demands, and services despite a rising number of consumers, such as portable gadgets in the network of edge. The volume of various linked objects has recently rapidly increased, that might interrupt operations and its overall quality, as well as cause network difficulties owing to the massive amount of data generated by connected items [139]. As a result, the edge servers must assure service flexibility at the fog layer by employing solutions such as server clustering and bandwidth allocation [32].

1.10.2   Issue of Trust

A system comprising of numerous intelligent devices and communication protocols has multiple nodes [20, 140, 141]. When a specific hub in an IoT network is conquered, it is relatively easy to bring the IoT system to a halt. Because boosting the high efficiency and reliability of the IoT network is one of the best techniques to dealing with the collapse issue, several academics are interested in the trust management (TM) concern in the IoT field [142]. Figure 1.11 illustrates a graphical picture of the loss of IoT services due by different link collapse, and it depicts an IoT system made up of humans, automobiles, warehouses, and computers.

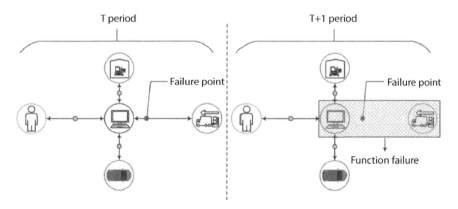

Figure 1.11 Node failures due to lack of IoT functions from [32].

Previous research suggests that TM might aid in the development of an IoT implementation services systems relying on the effective fusion of different sensor complex data, verified quality of service provided and robust ultimate customer secrecy and security protection [143]. To solve TM issues in IoT administration, Ning *et al.* [144] created an IoT network framework strategy based on four aspects, such as network/connections, data/information, system, and security of various application. Sun *et al.* [145] appears to employ a greater microlevel technique for resolving defensive managerial challenges when compared to previous studies. It extensively covered TM in the IoT from the viewpoints of monitoring data verification, light encryption techniques and procedures, and physically level fundamental guidelines. Moreover, according to Li and Zhou [146], IoT architecture, information processing security, and personal privacy protection are unavoidable problems during the building process. Liu and Wang [147] researched on the TM of IoT from three angles, including heterogeneous network model construction, confidence routing, and designing of TM and provided further detailed guidance for IoT security design.

1.10.3 Service Availability

The system has to provide service at all times and through any location with disintegration or distraction, that is taken into consideration a critical challenge for practical application scenarios. The difficulties raised includes such as streaming videos that requires a minimal time to process and reply, particularly in automobile EC, in which several elements like as vehicle motion, barriers, and so on might impair availability and reliability of system. In order to do so, several methods, such as system backups, prediction, and monitoring, must be implemented to ensure efficient service availability for the present and next generation of EC applications [32].

1.10.4 Security Challenges

The IoT is besides facing safety issues as it develops grow high. The earlier studies and investigation have detailed the safety of the IoT, as well as popular IoT attack vectors and remedies [148]. According to Harbi *et al.*, [149] have described that an IoT network's ability to connect a high number of devices increases the danger of a network attack. In most cases, these risks result in loss of data within system of IoT and in the worst-case scenario the loss of certain IoT functionality. The goal of these assaults is to terminate or gain unauthorized entry to the IoT structure, therefore attaining the goal of causing damage to the IoT system [150]. Attacks, such as snooze deficiency, hacking, and bogus data injection, have posed significant problems to IoT security. Following that, this section aims to offer some viable countermeasures to these assaults. Evaluating the efficiency of edge technologies' sustainable energy sources, including by installing solar recharging panels, is one way to address this issue. A phishing assault is when an attacker sends phishing URLs through mail or direct online chat windows [151].

Ultimate consumer is tempted to snap on given links, which are subsequently used by the hacker to collect confidential information such as usernames and passwords. Keeping clients aware of unexpected connections is one of the most efficient methods to address this problem. An attacker can inject fake information into an IoT node or edge device in a misleading information injection attack [152]. As a result of the fake data, one of several attacker's goals for damaging the IoT system is to force the IoT control system to issue wrong orders. To prevent a fake data injection attack, IoT nodes must be equipped with a data authenticity checking system.

1.10.5 Mobility Issues

Because most linked objects, such as smart phones, autos, and drones, are relatively mobile, regular connection failure between equipment and software is a key concern in the IoT of Things. This problem degrades the fog system's QoS and security; strong mobility device compatibility is a crucial problem that must be tackled in future network generations [153, 154]. As a result, several methods, such as link failure prediction to assure path stability, obstacle identification, and so on, must be proposed.

1.10.6 Architecture for IoT

SOA is a significant technique for combining network architectures or devices that has been effectively implemented in CC and automotive communication [155, 156]. The ability to build a dynamic multilayered SOA architecture based on individual business demands is critical for SOA to become extensively adopted in the IoT. According to the ITU, an IoT architecture should have five layers such as sensing, network, middle, access and its applications. Domingo [157] and Li et al., [158] have developed a three-part IoT layer framework, which covers perception, service and network layers. The essential four components of the IoT framework are things and devices, communication gateways, clouds and data centres, and applications and services [159].

The present mainstream IoT system consists of five sections based on the five independent sets, such as a device layer, transport layer, cloud layer, perception layer, and application layer. Multi-intelligent control mechanisms [160], cyberphysical systems [161], business model renewal [162], and smart manufacturing systems have been all used in the IoT design [163].

1.11 Conclusion

The IoT is constantly transforming our society, and it has been used in a variety of industries including digital medical services, smart cities, agriculture, environmental, and industrial production. Many scholars and practitioners have been studying the IoT since its inception. This current chapter provided an overview of the IoT's premise, supporting technologies, protocols, and applications, as well as recent research on various elements of the IoT over past 20 years. In addition, several of the problems and consequences are allied to the creation and deployment of IoT solutions were discussed. The relationship between IoT, cloud, and EC has also been discussed. Because of the increasing proliferation of IoT networks, EC is becoming a solution to the complex concerns and difficulties of controlling linked devices/sensors and the large computational resources they use. EC, in contrast to CC, will enable massive data computing and storage at the edge of the network that will be desired by future developments like as smart residential areas, smart cities, and smart vehicles, after a decade of development. A number of concerns and problems are linked to the development, and planning of IoT solutions was also covered.

Furthermore, this study examines the future issues that the IoT will encounter. The issues of trust and security, as well possible alternatives to these problems, are examined and debated in depth. The topics of trust and security, as well as possible alternatives to these problems, are thoroughly investigated and argued. Identifying how to construct a stronger trusted IoT system built on the appropriate integration of multisource content, approved quality of service, and robust privacy protection for ultimate customer are the concerns in the IoT trust sector. In terms of security, resume restriction attacks, cyber-attacks, and false data extrusion threats are all explored in full.

References

1. Bandyopadhyay, D. and Sen, J., Internet of Things: Applications and challenges in technology and standardization. *Wirel. Pers. Commun.*, *58*, 1, 49–69, 2011.
2. Santucci, G., From internet of data to Internet of Things, in: *International Conference on Future Trends of the Internet*, 2009, January, vol. 28, pp. 1–19.
3. Khanna, A. and Kaur, S., Internet of Things (IoT), applications and challenges: A comprehensive review. *Wirel. Pers. Commun.*, *114*, 2, 1687–1762, 2020.
4. Sha, K., Yang, T.A., Wei, W., Davari, S., A survey of EC-based designs for IoT security. *Digit. Commun. Netw.*, *6*, 2, 195–202, 2020.
5. Munín-Doce, A., Díaz-Casás, V., Trueba, P., Ferreno-González, S., Vilar-Montesinos, M., Industrial Internet of Things in the production environment of a Shipyard 4.0. *Int. J. Adv. Manuf. Technol.*, *108*, 1, 47–59, 2020.
6. Lakshmi, S.V., Janet, J., Rani, P.K., Sujatha, K., Satyamoorthy, K., Marichamy, S., Role and applications of IoT in materials and manufacturing industries–Review. *Mater. Today: Proc.*, *45*, 2925–2928, 2021.
7. Atzori, L., Iera, A., Morabito, G., The Internet of Things: A survey. *Comput. Networks*, *54*, 15, 2787–2805, 2010.
8. Lim, M.K., Wang, J., Wang, C., Tseng, M.L., A novel method for green delivery mode considering shared vehicles in the IoT environment. *Ind. Manage. Data Syst.*, *120*, 9, 1733–1757, 2020a.
9. Gubbi, J., Buyya, R., Marusic, S., Palaniswami, M., Internet of Things (IoT): A vision, architectural elements, and future directions. *Future Gener. Comput. Syst.*, *29*, 7, 1645–1660, 2013.
10. Su, J., Bai, Q., Sindakis, S., Zhang, X., Yang, T., Vulnerability of multinational corporation knowledge network facing resource loss: A super-network perspective. *Manage. Decis.*, *59*, 1, 84–103, 2019.
11. Kwak, J.Y., Cho, C., Shin, Y., Yang, S., IntelliTC: Intelligent inter-DC traffic controller for the internet of everything service based on fog computing. *IET Commun.*, *14*, 2, 193–205, 2020.
12. Gazis, V., A survey of standards for machine-to-machine and the Internet of Things. *IEEE Commun. Surv. Tutor.*, *19*, 1, 482–511, 2016.
13. Li, M., Internet of people. *Concurr. Comput.: Pract. Exp.*, *29*, 3, e4050, 2017.
14. Lim, M.K. and Jones, C., Resource efficiency and sustainability in logistics and supply chain management. *Int. J. Logist. Res. Appl.*, *20*, 1, 20–21, 2017.
15. Santoro, G., Vrontis, D., Thrassou, A., Dezi, L., The Internet of Things: Building a knowledge management system for open innovation and knowledge management capacity. *Technol. Forecasting Soc. Change*, *136*, 347–354, 2018.
16. Bouzembrak, Y., Klüche, M., Gavai, A., Marvin, H.J., Internet of Things in food safety: Literature review and a bibliometric analysis. *Trends Food Sci. Technol.*, *94*, 54–64, 2019.

17. Lim, M.K., Xiong, W., Lei, Z., Theory, supporting technology and application analysis of cloud manufacturing: A systematic and comprehensive literature review. *Ind. Manage. Data Syst.*, *120*, 8, 1585–1614, 2020, b.

18. Jiafu, S., Yu, Y., Tao, Y., Measuring knowledge diffusion efficiency in R&D networks. *Knowl. Manage. Res. Pract.*, *16*, 2, 208–219, 2018.

19. Tseng, M.L., Lim, M.K., Wu, K.J., Improving the benefits and costs on sustainable supply chain finance under uncertainty. *Int. J. Prod. Econ.*, *218*, 308–321, 2019.

20. Zhang, N., Yang, Y., Zheng, Y., Su, J., Module partition of complex mechanical products based on weighted complex networks. *J. Intell. Manuf.*, *30*, 4, 1973–1998, 2019.

21. Li, Y., Lim, M.K., Tan, Y., Lee, Y., Tseng, M.L., Sharing economy to improve routing for urban logistics distribution using electric vehicles. *Resour. Conserv. Recycl.*, *153*, 104585, 2020.

22. Al-Fuqaha, A., Guizani, M., Mohammadi, M., Aledhari, M., Ayyash, M., Internet of Things: A survey on enabling technologies, protocols, and applications. *IEEE Commun. Surv. Tutor.*, *17*, 4, 2347–2376, 2015.

23. Tu, M., Lim, M.K., Yang, M.F., IoT-based production logistics and supply chain system–Part 2: IoT-based cyber-physical system: A framework and evaluation. *Ind. Manage. Data Syst.*, *118*, 1, 96–125, 2018.

24. Sinha, A., Shrivastava, G., Kumar, P., Architecting user-centric internet of things for smart agriculture. *Sustain. Comput.: Inform. Syst.*, *23*, 88–102, 2019.

25. Manavalan, E. and Jayakrishna, K., A review of Internet of Things (IoT) embedded sustainable supply chain for industry 4.0 requirements. *Comput. Ind. Eng.*, *127*, 925–953, 2019.

26. Al-Turjman, F., Nawaz, M.H., Ulusar, U.D., Intelligence in the Internet of Medical Things era: A systematic review of current and future trends. *Comput. Commun.*, *150*, 644–660, 2020.

27. Muñuzuri, J., Onieva, L., Cortés, P., Guadix, J., Using IoT data and applications to improve port-based intermodal supply chains. *Comput. Ind. Eng.*, *139*, 105668, 2020.

28. Liu, K., Bi, Y., Liu, D., Internet of Things based acquisition system of industrial intelligent bar code for smart city applications. *Comput. Commun.*, *150*, 325–333, 2020.

29. Chahal, R.K., Kumar, N., Batra, S., Trust management in social Internet of Things: A taxonomy, open issues, and challenges. *Comput. Commun.*, *150*, 13–46, 2020.

30. Kiritsis, D., Closed-loop PLM for intelligent products in the era of the Internet of Things. *Comput.-Aided Des.*, *43*, 5, 479–501, 2011.

31. Deebak, B.D. and Al-Turjman, F., A hybrid secure routing and monitoring mechanism in IoT-based wireless sensor networks. *Ad Hoc Netw.*, *97*, 102022, 2020.

32. Laroui, M., Nour, B., Moungla, H., Cherif, M.A., Afifi, H., Guizani, M., Edge and fog computing for IoT: A survey on current research activities & future directions. *Comput. Commun.*, *180*, 210–231, 2021.

33. Shih, D.H., Sun, P.L., Yen, D.C., Huang, S.M., Taxonomy and survey of RFID anti-collision protocols. *Comput. Commun.*, *29*, 11, 2150–2166, 2006.

34. Juels, A., RFID security and privacy: A research survey. *IEEE J. Sel. Areas Commun.*, *24*, 2, 381–394, 2006.

35. Welbourne, E., Battle, L., Cole, G., Gould, K., Rector, K., Raymer, S., Borriello, G., Building the internet of things using RFID: The RFID ecosystem experience. *IEEE Internet Comput.*, *13*, 3, 48–55, 2009.

36. Zhang, S., Lin, Y., Liu, Q., Jiang, J., Yin, B., Choo, K.K.R., Secure hitch in location based social networks. *Comput. Commun.*, *100*, 65–77, 2017.

37. Mohandes, M.A., Mobile technology for socio-religious events: A case study of NFC technology. *IEEE Technol. Soc. Mag.*, *34*, 1, 73–79, 2015.

38. Beloglazov, A., Abawajy, J., Buyya, R., Energy-aware resource allocation heuristics for efficient management of data centers for cloud computing. *Future Gener. Comput. Syst.*, *28*, 5, 755–768, 2012.

39. Guo, H., Zheng, Y., Li, X., Li, Z., Xia, C., Self-healing group key distribution protocol in wireless sensor networks for secure IoT communications. *Future Gener. Comput. Syst.*, *89*, 713–721, 2018.

40. Wan, S., Gu, Z., Ni, Q., Cognitive computing and wireless communications on the edge for healthcare service robots. *Comput. Commun.*, *149*, 99–106, 2020.

41. Krčo, S., Pokrić, B., Carrez, F., Designing IoT architecture (s): A European perspective, in: *2014 IEEE World Forum on Internet of Things (WF-IoT)*, 2014, March, IEEE, pp. 79–84.

42. Morton, N.A. and Hu, Q., Implications of the fit between organizational structure and ERP: A structural contingency theory perspective. *Int. J. Inf. Manage.*, *28*, 5, 391–402, 2008.

43. Marston, S., Li, Z., Bandyopadhyay, S., Zhang, J., Ghalsasi, A., Cloud computing—The business perspective. *Decis. Support Syst.*, *51*, 1, 176–189, 2011.

44. Zissis, D. and Lekkas, D., Addressing cloud computing security issues. *Future Gener. Comput. Syst.*, *28*, 3, 583–592, 2012.

45. Mateescu, G., Gentzsch, W., Ribbens, C.J., Hybrid computing—Where HPC meets grid and cloud computing. *Future Gener. Comput. Syst.*, *27*, 5, 440–453, 2011.

46. Kumar, R., Singh, S.P., Lamba, K., Sustainable robust layout using Big Data approach: A key towards industry 4.0. *J. Cleaner Prod.*, *204*, 643–659, 2018.

47. Mell, P. and Grance, T., The NIST definition of cloud computing. *National Institute of Standards and Technology*, *53*, 6, 50, 2011.

48. Lee, K., Murray, D., Hughes, D., Joosen, W., Extending sensor networks into the cloud using amazon web services, in: *2010 IEEE International Conference on Networked Embedded Systems for Enterprise Applications*, 2010, November, IEEE, pp. 1–7.

49. Aitken, R., Chandra, V., Myers, J., Sandhu, B., Shifren, L., Yeric, G., Device and technology implications of the Internet of Things, in: *2014 Symposium on VLSI Technology (VLSI-technology): Digest of Technical Papers*, pp. 1–4, IEEE, 2014, June.

50. Alhakbani, N., Hassan, M.M., Hossain, M.A., Alnuem, M., A framework of adaptive interaction support in cloud-based Internet of Things (IoT) environment, in: *International Conference on Internet and Distributed Computing Systems*, pp. 136–146, Springer, Cham, 2014, September.

51. Botta, A., De Donato, W., Persico, V., Pescapé, A., Integration of cloud computing and Internet of Things: A survey. *Future Gener. Comput. Syst.*, 56, 684–700, 2016.

52. Dash, S.K., Mohapatra, S., Pattnaik, P.K., A survey on applications of wireless sensor network using cloud computing. *Int. J. Comput. Sci. Emerging Technol.*, 1, 4, 50–55, 2010.

53. Suciu, G., Vulpe, A., Halunga, S., Fratu, O., Todoran, G., Suciu, V., Smart cities built on resilient cloud computing and secure Internet of Things, in: *2013 19th International Conference on Control Systems and Computer Science*, pp. 513–518, IEEE, May 2013.

54. Malik, A. and Om, H., Cloud computing and Internet of Things integration: Architecture, applications, issues, and challenges, in: *Sustainable Cloud and Energy Services*, pp. 1–24, Springer, Cham, 2018.

55. Rao, B.P., Saluia, P., Sharma, N., Mittal, A., Sharma, S.V., Cloud computing for Internet of Things & sensing based applications, in: *2012 Sixth International Conference on Sensing Technology (ICST)*, pp. 374–380, IEEE, 2012, December.

56. Zaslavsky, A., Perera, C., Georgakopoulos, D., Sensing as a service and big data. *arXiv preprint arXiv:1301.0159*, 21–29, 2013.

57. Prati, A., Vezzani, R., Fornaciari, M., Cucchiara, R., Intelligent video surveillance as a service, in: *Intelligent Multimedia Surveillance*, pp. 1–16, Springer, Berlin, Heidelberg, 2013.

58. Cavalcante, E., Pereira, J., Alves, M.P., Maia, P., Moura, R., Batista, T., Pires, P.F., On the interplay of Internet of Things and Cloud Computing: A systematic mapping study. *Comput. Commun.*, 89, 17–33, 2016.

59. Neagu, G., Preda, Ş., Stanciu, A., Florian, V., A cloud-IoT based sensing service for health monitoring, in: *2017 E-Health and Bioengineering Conference (EHB)*, pp. 53–56, IEEE, 2017, June.

60. Ismail, L. and Materwala, H., Energy-aware VM placement and task scheduling in cloud-IoT computing: Classification and performance evaluation. *IEEE Internet Things J.*, 5, 5166–5176, 2018.

61. Almolhis, N., Alashjaee, A.M., Duraibi, S., Alqahtani, F., Moussa, A.N., The security issues in IoT-cloud: A review, in: *2020 16th IEEE International Colloquium on Signal Processing & Its Applications (CSPA)*, pp. 191–196, IEEE, 2020, February.

62. Shi, W., Cao, J., Zhang, Q., Li, Y., Xu, L., EC: Vision and challenges. *IEEE Internet Things J.*, 3, 5, 637–646, 2016.
63. Nour, B., Mastorakis, S., Mtibaa, A., Compute-less networking: Perspectives, challenges, and opportunities. *IEEE Network*, 34, 6, 259–265, 2020.
64. Satyanarayanan, M., Bahl, P., Caceres, R., Davies, N., The case for vm-based cloudlets in mobile computing. *IEEE Pervasive Comput.*, 8, 4, 14–23, 2009.
65. Chen, B., Wan, J., Celesti, A., Li, D., Abbas, H., Zhang, Q., EC in IoT-based manufacturing. *IEEE Commun. Mag.*, 56, 9, 103–109, 2018.
66. Yi, S., Li, C., Li, Q., A survey of fog computing: Concepts. *Applications and Issues*, in: *Proceedings of the 2015 Workshop on Mobile Big Data (Mobidata '15)*, 37–42, 2016, http://dx.doi.org/10.1145/2757384.2757397.
67. Pan, J. and McElhannon, J., Future edge cloud and EC for Internet of Things applications. *IEEE Internet Things J.*, 5, 1, 439–449, 2017.
68. Sittón-Candanedo, I., Alonso, R.S., Corchado, J.M., Rodríguez-González, S., Casado-Vara, R., A review of EC reference architectures and a new global edge proposal. *Future Gener. Comput. Syst.*, 99, 278–294, 2019.
69. Singh, D., Tripathi, G., Jara, A.J., A survey of Internet-of-Things: Future vision, architecture, challenges and services, in: *2014 IEEE World Forum on Internet of Things (WF-IoT)*, pp. 287–292, IEEE, 2014, March.
70. Lin, J., Yu, W., Zhang, N., Yang, X., Ge, L., On data integrity attacks against route guidance in transportation-based cyber-physical systems, in: *2017 14th IEEE Annual Consumer Communications & Networking Conference (CCNC)*, pp. 313–318, IEEE, 2017, a January.
71. Sha, K., Yang, T.A., Wei, W., Davari, S., A survey of EC-based designs for IoT security. *Digit. Commun. Netw.*, 6, 2, 195–202, 2020.
72. Chen, X., Jiao, L., Li, W., Fu, X., Efficient multi-user computation offloading for mobile-edge cloud computing. *IEEE/ACM Trans. Netw.*, 24, 5, 2795–2808, 2015.
73. Zorzi, M., Gluhak, A., Lange, S., Bassi, A., From today's Intranet of Things to a future Internet of Things: A wireless-and mobility-related view. *IEEE Wirel. Commun.*, 17, 6, 44–51, 2010.
74. Hank, P., Müller, S., Vermesan, O., Van Den Keybus, J., Automotive ethernet: In-vehicle networking and smart mobility, in: *2013 Design, Automation & Test in Europe Conference & Exhibition (DATE)*, pp. 1735–1739, IEEE, 2013, March.
75. Kyriazis, D., Varvarigou, T., White, D., Rossi, A., Cooper, J., Sustainable smart city IoT applications: Heat and electricity management & eco-conscious cruise control for public transportation, in: *2013 IEEE 14th International Symposium on" A World of Wireless, Mobile and Multimedia Networks"(WoWMoM)*, pp. 1–5, IEEE, 2013, June.
76. Somov, A., Dupont, C., Giaffreda, R., Supporting smart-city mobility with cognitive Internet of Things, in: *2013 Future Network & Mobile Summit*, pp. 1–10, IEEE, 2013, July.
77. Vermesan, O., Blystad, L.C., Hank, P., Bahr, R., John, R., Moscatelli, A., Smart, connected and mobile: Architecting future electric mobility ecosystems,

in: *2013 Design, Automation & Test in Europe Conference & Exhibition (DATE)*, pp. 1740–1744, IEEE, 2013, March.

78. He, W., Yan, G., Da Xu, L., Developing vehicular data cloud services in the IoT environment. *IEEE Trans. Ind. Inf.*, 10, 2, 1587–1595, 2014.

79. Jin, J., Gubbi, J., Marusic, S., Palaniswami, M., An information framework for creating a smart city through Internet of Things. *IEEE Internet Things J.*, 1, 2, 112–121, 2014.

80. Ma, X., Yu, H., Wang, Y., Wang, Y., Large-scale transportation network congestion evolution prediction using deep learning theory. *PLoS One*, 10, 3, e0119044, 2015.

81. Liu, J., Li, J., Zhang, L., Dai, F., Zhang, Y., Meng, X., Shen, J., Secure intelligent traffic light control using fog computing. *Future Gener. Comput. Syst.*, 78, 817–824, 2018.

82. Poslad, S., Ma, A., Wang, Z., Mei, H., Using a smart city IoT to incentivise and target shifts in mobility behaviour—Is it a piece of pie? *Sensors*, 15, 6, 13069–13096, 2015.

83. Gregoriades, A. and Sutcliffe, A., Simulation-based evaluation of an in-vehicle smart situation awareness enhancement system. *Ergonomics*, 61, 7, 947–965, 2018.

84. Singh, J. and Singh, K., Congestion control in vehicular ad hoc network: A review, in: *Next-Generation Networks*, pp. 489–496, 2018.

85. Komninos, N., Philippou, E., Pitsillides, A., Survey in smart grid and smart home security: Issues, challenges and countermeasures. *IEEE Commun. Surv. Tutor.*, 16, 4, 1933–1954, 2014.

86. Okay, F.Y. and Ozdemir, S., A fog computing based smart grid model, in: *2016 International Symposium on Networks, Computers and Communications (ISNCC)*, pp. 1–6, IEEE, 2016, May.

87. Muzakkir Hussain, M., Alam, M.S., Sufyan Beg, M.M., Feasibility of fog computing in smart grid architectures, in: *Proceedings of 2nd International Conference on Communication, Computing and Networking*, pp. 999–1010, Springer, Singapore, 2019.

88. Khanna, A. and Kaur, S., Internet of things (IoT), applications and challenges: A comprehensive review. *Wirel. Pers. Commun.*, 114, 2, 1687–1762, 2020.

89. Bonomi, F., Milito, R., Zhu, J., Addepalli, S., Fog computing and its role in the Internet of Things, in: *Proceedings of the First Edition of the MCC Workshop on Mobile Cloud Computing*, 13–16, 2012.

90. Sulthana, S., Thatiparthi, G., Gunturi, R.S., Cloud and intelligent based SCADA technology. *Int. J. Adv. Res. Comput. Sci. Electron. Eng. (IJARCSEE)*, 2, 3, 293–296, 2013.

91. Karnouskos, S. and De Holanda, T.N., Simulation of a smart grid city with software agents, in: *2009 Third UKSim European Symposium on Computer Modeling and Simulation*, pp. 424–429, IEEE, 2009, November.

92. Bressan, N., Bazzaco, L., Bui, N., Casari, P., Vangelista, L., Zorzi, M., The deployment of a smart monitoring system using wireless sensor and actuator networks, in: *2010 First IEEE International Conference on Smart Grid Communications*, pp. 49–54, IEEE, 2010, October.

93. Karnouskos, S., The cooperative Internet of Things enabled smart grid, in: *Proceedings of the 14th IEEE International Symposium on Consumer Electronics (ISCE2010)*, pp. 07–10, 2010, June.

94. Farhangi, H., The path of the smart grid. *IEEE Power Energy Mag.*, 8, 1, 18–28, 2009.

95. Jiang, J., Gao, L., Jin, J., Luan, T.H., Yu, S., Xiang, Y., Garg, S., Sustainability analysis for fog nodes with renewable energy supplies. *IEEE Internet Things J.*, 6, 4, 6725–6735, 2019.

96. Zheng, Q., Jin, J., Zhang, T., Li, J., Gao, L., Xiang, Y., Energy-sustainable fog system for mobile web services in infrastructure-less environments. *IEEE Access*, 7, 161318–161328, 2019.

97. Mano, L.Y., Faiçal, B.S., Nakamura, L.H., Gomes, P.H., Libralon, G.L., Meneguete, R.I., Ueyama, J., Exploiting IoT technologies for enhancing Health Smart Homes through patient identification and emotion recognition. *Comput. Commun.*, 89, 178–190, 2016.

98. Han, D.M. and Lim, J.H., Design and implementation of smart home energy management systems based on zigbee. *IEEE Trans. Consum. Electron.*, 56, 3, 1417–1425, 2010.

99. Chong, G., Zhihao, L., Yifeng, Y., The research and implement of smart home system based on Internet of Things, in: *2011 International Conference on Electronics, Communications and Control (ICECC)*, pp. 2944–2947, IEEE, 2011, September.

100. Jie, Y., Pei, J.Y., Jun, L., Yun, G., Wei, X., Smart home system based on IoT technologies, in: *2013 International Conference on Computational and Information Sciences*, pp. 1789–1791, IEEE, 2013, June.

101. Soliman, M., Abiodun, T., Hamouda, T., Zhou, J., Lung, C.H., Smart home: Integrating Internet of Things with web services and cloud computing, in: *2013 IEEE 5th International Conference on Cloud Computing Technology and Science*, vol. 2, pp. 317–320, IEEE, 2013, December.

102. Wang, M., Zhang, G., Zhang, C., Zhang, J., Li, C., An IoT-based appliance control system for smart homes, in: *2013 Fourth International Conference on Intelligent Control and Information Processing (ICICIP)*, pp. 744–747, IEEE, 2013, June.

103. Rahmani, A.M., Liljeberg, P., Preden, J.S., Jantsch, A. (Eds.), *Fog computing in the Internet of Things: Intelligence at the edge*, Switzerland, Springer, 2017.

104. Wang, X., Gu, B., Ren, Y., Ye, W., Yu, S., Xiang, Y., Gao, L., A fog-based recommender system. *IEEE Internet Things J.*, 7, 2, 1048–1060, 2019.

105. Castellani, A.P., Gheda, M., Bui, N., Rossi, M., Zorzi, M., Web services for the Internet of Things through CoAP and EXI, in: *2011 IEEE International Conference on Communications Workshops (ICC)*, pp. 1–6, 2011, June.

106. Oliveira, L.M. and Rodrigues, J.J., Wireless sensor networks: A survey on environmental monitoring. *J. Commun.*, 6, 2, 143–151, 2011.

107. Jia, X., Feng, Q., Fan, T., Lei, Q., RFID technology and its applications in Internet of Things (IoT), in: *2012 2nd International Conference on Consumer Electronics, Communications and Networks (CECNet)*, pp. 1282–1285, IEEE, 2012, April.

108. Lazarescu, M.T., Design of a WSN platform for long-term environmental monitoring for IoT applications. *IEEE J. Emerging Sel. Top. Circuits Syst.*, 3, 1, 45–54, 2013.

109. Zhao, J., Zheng, X., Dong, R., Shao, G., The planning, construction, and management toward sustainable cities in China needs the Environmental Internet of Things. *Int. J. Sustain. Dev. World Ecol.*, 20, 3, 195–198, 2013.

110. Fang, S., Da Xu, L., Zhu, Y., Ahati, J., Pei, H., Yan, J., Liu, Z., An integrated system for regional environmental monitoring and management based on Internet of Things. *IEEE Trans. Ind. Inform.*, 10, 2, 1596–1605, 2014.

111. Ahmadi, H., Arji, G., Shahmoradi, L., Safdari, R., Nilashi, M., Alizadeh, M., The application of Internet of Things in healthcare: A systematic literature review and classification. *Univers. Access Inf. Soc.*, 18, 4, 837–869, 2019.

112. Kashani, M.H., Madanipour, M., Nikravan, M., Asghari, P., Mahdipour, E., A systematic review of IoT in healthcare: Applications, techniques, and trends. *J. Netw. Comput. Appl.*, 192, 103164, 2021.

113. Gia, T.N., Dhaou, I.B., Ali, M., Rahmani, A.M., Westerlund, T., Liljeberg, P., Tenhunen, H., Energy efficient fog-assisted IoT system for monitoring diabetic patients with cardiovascular disease. *Future Gener. Comput. Syst.*, 93, 198–211, 2019.

114. Mahmoud, M.M., Rodrigues, J.J., Saleem, K., Al-Muhtadi, J., Kumar, N., Korotaev, V., Towards energy-aware fog-enabled cloud of things for healthcare. *Comput. Electr. Eng.*, 67, 58–69, 2018.

115. Tuli, S., Basumatary, N., Gill, S.S., Kahani, M., Arya, R.C., Wander, G.S., Buyya, R., HealthFog: An ensemble deep learning based Smart Healthcare System for Automatic Diagnosis of Heart Diseases in integrated IoT and fog computing environments. *Future Gener. Comput. Syst.*, 104, 187–200, 2020.

116. Suma, N., Samson, S.R., Saranya, S., Shanmugapriya, G., Subhashri, R., IOT based smart agriculture monitoring system. *Int. J. Recent Innov. Trends Comput. Commun.*, 5, 2, 177–181, 2017.

117. Kumari, M., An overview of future IoT systems: Applications, challenges, and future trends, in: *Electronic Devices and Circuit Design: Challenges and Applications in the Internet of Things*, pp. 1–17, 2022.

118. Patil, K.A. and Kale, N.R., A model for smart agriculture using IoT, in: *2016 International Conference on Global Trends in Signal Processing, Information Computing and Communication (ICGTSPICC)*, pp. 543–545, IEEE, 2016, December.

119. Khoa, T.A., Man, M.M., Nguyen, T.Y., Nguyen, V., Nam, N.H., Smart agriculture using IoT multi-sensors: A novel watering management system. *J. Sens. Actuator Netw.*, 8, 3, 45, 2019.

120. Naresh, M. and Munaswamy, P., Smart agriculture system using IoT technology. *Int. J. Recent Technol. Eng.*, 7, 5, 98–102, 2019.

121. Kumar, K.N., Pillai, A.V., Narayanan, M.B., Smart agriculture using IoT. *Mater. Today: Proc.*, 2021. https://doi.org/10.1016/j.matpr.2021.02.474

122. Kuo, C.J., Ting, K.C., Chen, Y.C., Yang, D.L., Chen, H.M., Automatic machine status prediction in the era of Industry 4.0: Case study of machines in a spring factory. *J. Syst. Archit.*, 81, 44–53, 2017.

123. Bauza, M.B., Tenboer, J., Li, M., Lisovich, A., Zhou, J., Pratt, D., Knebel, R., Realization of Industry 4.0 with high speed CT in high volume production. *CIRP J. Manuf. Sci. Technol.*, 22, 121–125, 2018.

124. Santos, M.Y., e Sá, J.O., Andrade, C., Lima, F.V., Costa, E., Costa, C., Galvão, J., A big data system supporting bosch braga industry 4.0 strategy. *Int. J. Inf. Manage.*, 37, 6, 750–760, 2017.

125. Lin, C., He, D., Huang, X., Choo, K.K.R., Vasilakos, A.V., BSeIn: A blockchain-based secure mutual authentication with fine-grained access control system for industry 4.0. *J. Netw. Comput. Appl.*, 116, 42–52, 2018.

126. Dinardo, G., Fabbiano, L., Vacca, G., A smart and intuitive machine condition monitoring in the Industry 4.0 scenario. *Measurement*, 126, 1–12, 2018.

127. Peddireddy, D., Fu, X., Wang, H., Joung, B.G., Aggarwal, V., Sutherland, J.W., Jun, M.B.G., Deep learning based approach for identifying conventional machining processes from CAD data. *Proc. Manuf.*, 48, 915–925, 2020.

128. Kim, H., Yeo, C., Lee, I.D., Mun, D., Deep-learning-based retrieval of piping component catalogs for plant 3D CAD model reconstruction. *Comput. Ind.*, 123, 103320, 2020.

129. Anbalagan, A. and Moreno-Garcia, C.F., An IoT based industry 4.0 architecture for integration of design and manufacturing systems. *Mater. Today: Proc.*, 46, 7135–7142, 2021.

130. Kusiak, A., Smart manufacturing. *Int. J. Prod. Res.*, 56, 1-2, 508–517, 2018.

131. Yao, X. and Lin, Y., Emerging manufacturing paradigm shifts for the incoming industrial revolution. *Int. J. Adv. Manuf. Technol.*, 85, 5, 1665–1676, 2016.

132. Srinivasan, M., Prince, E., Padmanabhan, R., IoT architecture for advanced manufacturing technologies. *Mater. Today: Proc.*, 22, 2359–2365, 2020.

133. Stanford, V., Pervasive computing goes the last hundred feet with RFID systems. *IEEE Pervasive Comput.*, 2, 2, 9–14, 2003.

134. Johnson, D., RFID tags improve tracking, quality on Ford line in Mexico. *Control Eng.*, 49, 11, 16–16, 2002.

135. Zhekun, L., Gadh, R., Prabhu, B.S., Applications of RFID technology and smart parts in manufacturing, in: *International Design Engineering Technical Conferences and Computers and Information in Engineering Conference*, vol. 46970, pp. 123–129, 2004, January.

136. Öztayşi, B., Baysan, S., Akpinar, F., Radio frequency identification (RFID) in hospitality. *Technovation*, *29*, 9, 618–624, 2009.
137. Budak, E., Catay, B., Tekin, I., Yenigun, H., Abbak, M., Drannikov, S., Simsek, O., Design of an RFID-based manufacturing monitoring and analysis system, in: *2007 1st Annual RFID Eurasia*, pp. 1–6, IEEE, 2007, September.
138. Asif, Z., Integrating the supply chain with RFID: A technical and business analysis. *Commun. Assoc. Inf. Syst.*, *15*, 1, 24, 2005.
139. Mouradian, C., Naboulsi, D., Yangui, S., Glitho, R.H., Morrow, M.J., Polakos, P.A., A comprehensive survey on fog computing: State-of-the-art and research challenges. *IEEE Commun. Surv. Tutor.*, *20*, 1, 416–464, 2017.
140. Ding, L., Hu, B., Ke, C., Wang, T., Chang, S., Effects of IoT technology on gray market: An analysis based on traceability system design. *Comput. Ind. Eng.*, *136*, 80–94, 2019.
141. Zhang, N., Yang, Y., Wang, J., Li, B., Su, J., Identifying core parts in complex mechanical product for change management and sustainable design. *Sustainability*, *10*, 12, 4480, 2018.
142. Saied, Y.B., Olivereau, A., Zeghlache, D., Laurent, M., Trust management system design for the Internet of Things: A context-aware and multi-service approach. *Comput. Secur.*, *39*, 351–365, 2013.
143. Yan, Z., Zhang, P., Vasilakos, A.V., A survey on trust management for Internet of Things. *J. Netw. Comput. Appl.*, *42*, 120–134, 2014.
144. Ning, H., Liu, H., Yang, L.T., Cyberentity security in the Internet of Things. *Computer*, *46*, 4, 46–53, 2013.
145. Sun, W., Cai, Z., Li, Y., Liu, F., Fang, S., Wang, G., Security and privacy in the medical Internet of Things: A review. *Secur. Commun. Netw.*, *2018*, 1–9, 2018.
146. Li, H. and Zhou, X., Study on security architecture for Internet of Things, in: *International Conference on Applied Informatics and Communication*, pp. 404–411, Springer, Berlin, Heidelberg, 2011, August.
147. Liu, Y. and Wang, K., Trust control in heterogeneous networks for Internet of Things, in: *2010 International Conference on Computer Application and System Modeling (ICCASM 2010)*, vol. 1, pp. V1–632, IEEE, 2010, October.
148. Lin, J., Yu, W., Zhang, N., Yang, X., Zhang, H., Zhao, W., A survey on Internet of Things: Architecture, enabling technologies, security and privacy, and applications. *IEEE Internet Things J.*, *4*, 5, 1125–1142, 2017b.
149. Harbi, Y., Aliouat, Z., Harous, S., Bentaleb, A., Refoufi, A., A review of security in Internet of Things. *Wirel. Pers. Commun.*, *108*, 1, 325–344, 2019.
150. Husain, A.J. and Mohamed, M.A., IMBF-counteracting denial-of-sleep attacks in 6LowPAN based Internet of Things. *J. Inf. Sci. Eng.*, *35*, 2, 361–374, 2019.
151. Spaulding, J. and Mohaisen, A., Defending Internet of Things against malicious domain names using D-FENS, in: *2018 IEEE/ACM Symposium on EC (SEC)*, pp. 387–392, IEEE, 2018, October.

152. Liu, X., Qian, C., Hatcher, W.G., Xu, H., Liao, W., Yu, W., Secure Internet of Things (IoT)-based smart-world critical infrastructures: Survey, case study and research opportunities. *IEEE Access*, *7*, 79523–79544, 2019.

153. Aggarwal, S. and Kumar, N., Path planning techniques for unmanned aerial vehicles: A review, solutions, and challenges. *Comput. Commun.*, *149*, 270–299, 2020.

154. Wang, R., Cao, Y., Noor, A., Alamoudi, T.A., Nour, R., Agent-enabled task offloading in UAV-aided mobile EC. *Comput. Commun.*, *149*, 324–331, 2020.

155. Da Silva, F.S.T., da Costa, C.A., Crovato, C.D.P., da Rosa Righi, R., Looking at energy through the lens of Industry 4.0: A systematic literature review of concerns and challenges. *Comput. Ind. Eng.*, *143*, 106426, 2020.

156. Paulraj, D., Swamynathan, S., Madhaiyan, M., Process model-based atomic service discovery and composition of composite semantic web services using web ontology language for services (OWL-S). *Enterp. Inf. Syst.*, *6*, 4, 445–471, 2012.

157. Domingo, M.C., An overview of the Internet of Things for people with disabilities. *J. Netw. Comput. Appl.*, *35*, 2, 584–596, 2012.

158. Li, Q., Wang, Z.Y., Li, W.H., Li, J., Wang, C., Du, R.Y., Applications integration in a hybrid cloud computing environment: Modelling and platform. *Enterp. Inf. Syst.*, *7*, 3, 237–271, 2013.

159. Hossain, M.S. and Muhammad, G., Cloud-assisted industrial Internet of Things (IIoT)–enabled framework for health monitoring. *Comput. Networks*, *101*, 192–202, 2016.

160. Hadipour, M., Derakhshandeh, J.F., Shiran, M.A., An experimental setup of Multi-Intelligent Control System (MICS) of water management using the Internet of Things (IoT). *ISA Trans.*, *96*, 309–326, 2020.

161. Silva, E.M. and Jardim-Goncalves, R., Cyber-physical systems: A multi-criteria assessment for Internet-of-Things (IoT) systems. *Enterp. Inf. Syst.*, *15*, 3, 332–351, 2021.

162. Rocha, C., Narcizo, C.F., Gianotti, E., Internet of management artifacts: Internet of Things architecture for business model renewal, in: *Emerging Issues and Trends in Innovation and Technology Management*, pp. 297–316, 2022.

163. Jeong, S., Na, W., Kim, J., Cho, S., Internet of Things for smart manufacturing system: Trust issues in resource allocation. *IEEE Internet Things J.*, *5*, 6, 4418–4427, 2018.

Fourth Industrial Revolution: Industry 4.0

Maheswari Rajamanickam[1], Elizabeth Nirmala John Gerard Royan[1],
Gowtham Ramaswamy[1], Manivannan Rajendran[2] and Vaishnavi Vadivelu[1*]

[1]*Department of Management Studies, Kongu Engineering College, Erode,
Tamil Nadu, India*
[2]*Department of Mechanical Engineering, Kongu Engineering College, Erode,
Tamil Nadu, India*

Abstract

Industry 4.0 (I4.0) is the latest trend that has been implemented by many companies to increase their production and process efficiency. By including many automation techniques like artificial intelligence (AI), Internet of Things (IoT), cloud computing (CC), cyber-physical systems (CPS), cognitive computing, and much more, the collective version of these new techniques in a single company called smart factory. Many firms are trying to implement these digital techniques since this era's market is competitive, which has more advantage over the company which has not implemented I4.0. The functions are not only implemented in the production unit but also included in the complete chain starting from raw material suppliers to customers and also in service industries. By involving I4.0, factories reduce a lot of unwanted expenses and improve the quality of the products. This leads to innovative products, shift flexibility, effective use of resources, and increase customer satisfaction. Each firm needs to develop its own I4.0 structural methods according to its process flow and need. Manufacturing resources are intimately connected to data and pieces of information exchange, and quality and process control are maintained at all times. Every function has its software specially designed to carry out the actions and linked with the main functional system. The integrated process combining many roles—employing data mining to handle massive volumes of data and a decision support system to generate varying levels of the relevance of criteria with predictive analysis to identify when some criteria should be prioritized over others—is a practical application that has been employed in many industries, which has been discussed in some industrial case

Corresponding author: vaishnacsca@gmail.com

R. Rajasekar, C. Moganapriya, P. Sathish Kumar and M. Harikrishna Kumar (eds.) Integration of Mechanical and Manufacturing Engineering with IoT: A Digital Transformation, (41–84) © 2023 Scrivener Publishing LLC

study. Various applications of individual techniques are elaborated with examples and their further scope in multiple fields. I4.0 has already proved its potential in each field where it has already been implemented and people started to strive for faster growth.

Keywords: Industry 4.0, automation techniques, AI, IoT, CC, CPS, cognitive computing, smart factory, integrated functions

2.1   Introduction

Industry 4.0 (I4.0) says that the latest trend in technological automation and data sharing, which includes Cloud Computing (CC), Internet of Things (IoT), Cognitive Computing, and Cyber-Physical Systems (CPS) as well as, the creation of smart factory. In this globalization era, industrial firms are constantly under pressure to innovate, increase their competitiveness,and outperform their worldwide competitors [1].

One of their most potent partners in these endeavors is digital technology, which can help them to increase automation, decrease error-prone procedures, improve proactivity, streamline corporate operations, make processes knowledge-intensive, reduce costs, promote smartness and do more with less [2]. Furthermore, technology acceleration trends provide them with a plethora of chances for innovating in their processes and altering their operations in a way that results in a disruptive paradigm shift in their operations rather than marginal productivity gains. This is why, as part of a broader and strategic digital transformation plan, many industrial firms are actively investing in the digitization of their processes [3].

2.1.1   Global Level Adaption

In the year 2011, the government of Germany introduced a new concept known as I4.0 and it is regarded as the 4th industrial revolution are depicted in Figure 2.1. The global recession has shifted the focus on the industrial sector in recent years, with a focus on value-added products. A firmthat follows the strategy of relocating operations in search of lower-cost labor is now fighting for its success. The manufacturing firm in Germany used a strategy that played a crucial part in this transformation, beginning its steps to retain and promote the country's position as a "forerunner" in the industry [4]. With the introduction of the buzzword "I4.0," enormous hope appears to meet the latest challenges in the production process. Using its technologies, the concept I4.0 is facilitating and strengthening this style,

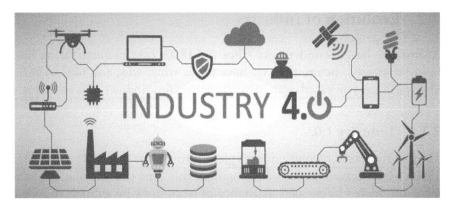

Figure 2.1 Integrated I4.0.

transforming people's lifestyles, establishing new business models and production methods, and revitalizing the new form of industry is digital transformation.

I4.0's goal is to operate with a higher level of automation to achieve greater efficiency and operational productivity by integrating the physical and virtual worlds. The traditional industry will benefit from computerization and interconnection [5]. I4.0 can be described as the production of knowledge integration and Cyber-Physical Systems (CPS) based on heterogeneous data. Also, it can be stated that an interpretational process of algorithms can be correlated, Big Data (BD) and high technology, such as the IoT and industrial automation, CC, cyber security (CS), services (IoS), or Intel. I4.0 is defined by Martin and Schaffer as "the communication between machines based on the real-time intelligent flow of workpieces machine-by-machine in a factory."

I4.0 will enable production to become smarter and more adaptable in this environment and to make better decisions in order to solve the problem by utilizing flexible and collaborative systems. It contributes to the advancement of the industrial scenario by focusing on the creation of a smart product, its processes, and its procedures. Companies are expected one raise the digitalization level by collaborating with customers and suppliers in digital ecosystems [6]. Since the I4.0 boom, researchers have seen a variety of approaches to the concept of I4.0; nevertheless, the general public may be perplexed due to a lack of understanding in this field. Clarification of I4.0-related concepts and technology is required [7, 8]. That is based on the smart factory (SF) idea. The SF idea incorporates elements such as systems integration, IoT, IoS, and the cyber-physical production system (CPPS), which is made up of numerous linked CPS.

2.2   Evolution of Industry

At first, we have to understand since the 1800s how manufacturing has been progressed. Before that, we have to see what, why, and how of I4.0. The world has witnessed or is currently seeing four industrial revolutions.

2.2.1   Industry 1.0

The first industrial revolution takes place through the time of the 1700s to 1800s. Manufacturing evolved throughout this time period from manual labor conducted by people's assistance of work done by animals to more optimal kind of labor utilizingsteam-powerkit and waterperformed by people with some additional sorts of mechanical kit.

2.2.2   Industry 2.0

With adequate use of electricity in industries during the early 19th century, the globe started a second industrial revolution. Manufacturers were able to boost production only after the arrival of electricity. During this time, mass manufacturing methods were adopted as a means of increasing productivity.

2.2.3   Industry 3.0

The third industrial revolution began to emerge in the late 1950s, the production ofelectronics like computersbegan to increase when the manufacturers began to cultivate the technology. Manufacturers began to move their focus away from analogy and technology based on machinery and towards the digitalized technologies and automation of software during this time.

2.2.4   Industry 4.0 (or) I4.0

The fourth industrial revolution is also known as I4.0, it has evolved in the recent 10 years. With the help of interconnection via IoT, factual moment data can be accessed, and the preface of CPS and the I4.0 pushes the emphasis on digital conversion from the recent 10 years to a next new altitude [9, 10]. I4.0 takes a holistic, wide-ranging, and interconnected method to fabricate. This denotes the real and digital worlds, enabling improved cooperation and access across people, vendors, products, and departments [11].

I4.0, help business owners to have a deep understanding and control over their business, procedure, improve productivity in real time.

Manufacturing has undergone various changes, at first by using the mechanical production and steam power, the second is the use of bulk production and electricity, and the third is the use of growing automation of manufacturing processes and information technology (IT). It has a connection towards the present trend, service orientation, advanced resources, and processing technology, and mutually sophisticated manufacturing networks (manufacturing the advanced devices inhibited by computers that combine them into a physical-digital environment) are driving the change in the era is called as I4.0 [12].

This shift affects the entire value chain, from raw materials through end-of-life recycling, also including business and its supportive services (e.g., sales, supply chain). Virtualization of manufacturing processes by connecting sensor data (byinspecting physical existence) with virtual plant and assumed prototype, centralized decision making, real-time potential to accumulate and analyze data and give insights, and flexible adaptation to changes are all examples of I4.0 design principles [1]. As anoutcome, I4.0 will open with innovative business models, offer solutions, and product prospects. As per the manufacturers' assessment whether they should pursue an I4.0 strategy, the change will test their ability to adapt at some time in the future [13]. By self-adjusting and adapting the supply chain, these influence the exponentially expanding amounts of data from sensors and connected devices in the operating environment, suppliers, and the distribution network will all provide issues [12]. Simultaneously, issues linked to data security will emerge as a result of greater connectivity [14].

2.3   Basic IoT Concepts and the Term Glossary

Maybe there are hundreds of key aspects and concepts associated with IoT and I4.0, but we have only 12 key terms and phrases, which understand before deciding either to invest or not in I4.0.

- (ERP) Enterprise Resource Planning: Tools for managing business processes that can be carried out to manage information throughout the organization.
- IoT: The Internet of Things (IoT) denotes the relationship between physical items like machinery and sensors and the internet.

- IIoT: Industrial Internet of Things (IIoT) denotes the connection between the people, data, and machines in the context of production.
- Big data analysis: Structured or unstructured data of a large set can be compiled, analyzed, sorted, and saved to discover relationships, trends, patterns, and opportunities.
- Artificial Intelligence (AI): Artificial intelligence refers to a computer›s competence to do tasks and make decisions that previously required some phase of individual intellect [15].
- M2M: M2M is an expansion of machine-to-machine communication and communication among two machines over a wireless or wired network [16].
- Digitization: Converting different types of data into digital format by the process of collecting and transforming the various information.
- Smartfactory: By taking the technology, solutions, and method in I4.0 to invest in the smart factory.
- Machine learning: The computer's ability to learn on its own in order to improve is called machine learning. Without being excessively directed or programmed to do so, regards to AI.
- Cloud computing (CC): Utilization of network for remote servers housed on the internet for processing data by the activities of storing, managing, and processing the data is called cloud computing.
- Real-time data processing (RTDP): Obtaining real-time or near–real-time results. RTDP refers to the ability of computer systems and devices to process data endlessly and autonomously.
- Ecosystem: An ecosystem in the manufacturing process refers to the ability of our completeprocess to be connected, including inventory and preparation, financials, customer connections, supply chain management (SCM), and manufacturing execution.
- Cyber-physical systems (CPS): CPS, also well known as cyber manufacturing, are an I4.0-enabled manufacturing environment that enables real-time data compilation, analysis, and transparency across all parts of a manufacturing operation [7, 17]. Now that we have a greater understanding of several key characteristics related to I4.0, we have learned more about how smart manufacturing may change the way we run and expand our firm.

2.4   Industrial Revolution

The Industrial Revolution is a term that implies a period of time when our society and economy were dramatically altered. In the context of a "revolution," which essentially means a swift and profound change, the term "development" may appear to imply some lag, yet there is little doubt that enormous changes occurred in a relatively short time. Small-scale workshops and craft studios were displaced by industries. The first factories to realize the new dawns were textile and pottery companies, and a new system of canals and railway lines facilitated efficient distribution. It was the start of a boom for both industries as they transitioned from industrious to industrial.

We can discern stages of four continuous processes known as the Industrial Revolution, starting with the 1st mechanical loom in 1784, exactly 230 years ago. That is how we are looking at it right now. The first "acceleration" happened around the end of the 18th century when mechanical industry based on water and steam was introduced. The advent of the conveyor belt and mass manufacturing, as well as the Henry Ford and Frederick Taylor, are associated with Second Industrial Revolution, which started to turn during the 20th century. The third option is to use electronics and information technology to digitally automate production.

2.4.1   I4.0 Core Idea

A trip through the four industrial revolutions might be symbolized by the steps in the growth of manufacturing industries, from manual labor to the idea of I4.0. The evolution is depicted in Figure 2.2. The digitalization of the 3rd industrial revolution is marked by the advent of microelectronics and computerization. This promotes stretchy production in manufacturing, where a wide range of items is made on flexible production lines using programmable machinery. However, such production systems lack flexibility in terms of production quantity.

The advancement of Information and Communication Technology (ICT) has ignited the fourth industrial revolution, which is now underway [19]. Smart automation of CPS, IoT, AR, BD and analytics, simulation, autonomous robots, IoS, CC, system integration, additive manufacturing, cyber security, and other technologies are all part of its technological foundation, which is supported by decentralized control and advanced connectivity (IoT functionalities). As a result of this new technology for industrial production systems, classic hierarchical automatic systems have

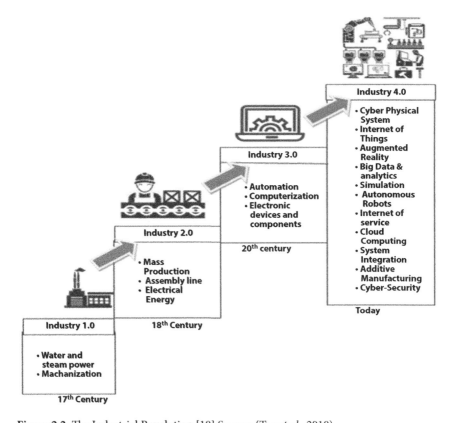

Figure 2.2 The Industrial Revolution [18] Source: (Tay *et al.*, 2018).

been rearranged into self-organizing cyberphysical production systems, allowing for flexible mass custom production and quality flexibility [20].

2.4.2   Origin of I4.0 Concept

It is no wonder that the I4.0 concept was established in Germany, which has one of the world's most competitive manufacturing industries and is even a global head in the industrial equipmentbusiness. I4.0 is a deliberate project spearheaded by the German regime, which has a long history of advancing technology.

I4.0 is a deliberate initiative spearheaded by the German regime, which has a long history of promoting technological advancement. In this sense, I4.0 can be considered as an attempt to maintain Germany's spot as one of the world's leading manufacturers of machinery and automobiles.

The basic impression was first introduced in 2011 at the Hannover Messe.

- Internet and IoT availability and utilization,
- Enterprise-wide integration of technology and business processes,
- Digital mapping and virtualization of the real world,
- A "smart" factory with "smart" industrial production procedures and "smart" products,
- I4.0 appears to be a viable method of reducing industrial production costs.

According to certain reports, I4.0 factories might reduce:

- Production expenses by 10% to 30%,
- Logistic costs by 10% to 30%, and
- Quality Management costs by 10% to 20%.

Adoption I4.0 will

- A faster time to market for innovative products,
- Increased customer responsiveness,
- Custom mass production without significantly increasing total production costs,
- A more flexible and friendly working environment, and
- A more efficient use of natural resources and energy are all advantages and reason forth is concept [20].

2.5   Industry

2.5.1   Manufacturing Phases

Manufacturer standards are currently undergoing significant changes. A faster reaction time is sought with a growing number of variants and a smaller batch size, resulting in product individualization. In firms that do not use cutting-edge technology, these standards are difficult to meet. Technology 4.0 ushers in a new era of intelligent factories producing custom-made things. The important technological developments that makeup technology 4.0 are as follows:

A new communication system that links the digital and physical worlds by allowing machines, items in various stages of production, systems, and people (each with their own Internet Protocol address) to exchange digital information via the internet protocol. Sophisticated human-machine interfaces; direct device connection.

- Intelligent sensors with built-in systems for individual recognition, data processing, and communication; incremental manufacturing technologies, such as 3D printing, are carried out in stages, and breakthrough changes are triggered by the scale of their application, synergy, integration, and development dynamics.
- Simulation approaches for the functioning of real items in virtual representations, based on data provided and processed in real time, allowing the configuration of production processes to be tested and optimized before making physical modifications; Simulation techniques for the operation of real objects in their virtual representations, depend on data provided and processed in real time, allowing to analyze and optimize manufacturing process configurations before introducing them; data processing in the cloud or fog, with millisecond response dynamics; analytics of large data sets on all aspects of product development and production;
- A new generation of robots capable of interacting with the environment and other robots while adapting to changing environment and requirements; augmented reality systems for device design and maintenance;
- Cyber security solutions allow for secure, dependable communication and information sharing. Access to systems and devices, as well as identification and management [22].

2.5.2 Existing Process Planning vs. I4.0

Standard, traditional Process Planning (PP) is exclusively dependent on the expertise and familiarity of a single individual or a group of work force in specific small- and mid-size enterprises (SMEs) in Croatia. These are technical experts who have gained their expertise through experience rather than formal schooling. They are educated using outdated methods that do not incorporate the concept of lifelong learning or the use of cutting-edge technology like Computer-aided process planning (CAPP) [23]. Because such people are difficult to persuade of the obvious need for change, this just illustrates the underlying issue and the obstacles that organizations would face by implementing I4.0. When big investments in innovative technology and the digitization process are required from the start, opposition grows [9].

The current features must be fully understood, assessed, and reviewed in order to meet the demand for change. I4.0 is defined as a digital and automated revolution of the entire organization, as well as the production process. I4.0 will soon be accepted by large global firms that use continuous improvement concepts and have high R&D standards, making them even more competitive in the market. SMEs that are still in the early stages of development will fall farther behind and will be unable to keep up with market changes and expectations.

As a result, they must develop their own I4.0 execution strategy as soon as possible. The new integrated factory's built-up resources are tightly linked to data and information interchange, with even quality and process control. The business network of the organization, as well as high-tech contact with consumers and suppliers, would be harmed. Smart products are becoming more widespread because they can transmit information and expertise from users/customers back to the manufacturing system, which can then be analyzed and optimized. Customers, who are at the centre of every manufacturing company's attention, may quickly and effectively participate in the process, build bespoke goods, and provide suggestions [24]. The structural dynamics control (SDC) technique can be used to resolve dynamic production and supply chain scheduling. The amount of information services required to keep physical systems running, as well as how these services will be scheduled during the planning stage and dynamically rescheduled throughout the implementation phase.

The information and expertise can be programmed into software that makes decisions basedon certain combinations and other technological requirements. When only one technology process is employed, this is acceptable, but what if the company produces a variety of products that require many production technologies?

The underlying principle is well known, and it can be used, but can it be solved in the same way as data grows larger and procedures become more complicated? The manufacturing process will become cloud manufacturing as a result of ongoing virtualization. Previously, only the hardware was used; now, the significance of contact networks and even virtual reality is growing. As a result, the term "smart product" has been popularized as a useful tool for getting reliable consumer feedback. Can the product, on the other hand, carry knowledge and make decisions at the level of PP and scheduling? This is an intriguing subject with a lot of potentials.

2.5.3   Software for Product Planning—A Link Between Smart Products and the Main System ERP

The term "smart products" was used to refer to information about client requests for customization, as well as user feedback. At all three stages of product design, development, and manufacturing, the goal is to integrate data collected by "smart products" into automated knowledge databases. Three of these are Process Planning (PP), operation sequencing, and scheduling (Deloitte, 2015). Following the product development part, parameters for the technology must be determined. CAPP, which serves as a bridge between CAD and CAM, has previously been used to automate PP (despite the many limitations and challenges that user's face).

With the surge in popularity of mass customization, a type of CAPP might be used, but given current technological developments, CAPP's generative approach is the best alternative. The purpose of the generative technique is to build a new progression plan for a given product using the knowledge database.

CAPP may be used to sequence and schedule tasks utilizing the same information, making it more extensively applicable in the I4.0 context. The CAD model generates the necessary code to define the technology. The new, bespoke technique will describe the code using a step model, reducing the need for a diagram. Specialized 3D modelling software is required. Despite the fact that generative CAPP-based software has been developed and is now in use, improvements and integration with other areas of the manufacturing process are required for I4.0. The supply chain and the processCAPP software is a type of "product planning" software that assists in the development of new products.

A few control mechanisms will be incorporated in the process to ensure that the new system is aware of the present system's limits. Advanced mathematical models and algorithms enable the identification of the product's surfaces and distinctive geometrical qualities in order to organize the method. Preliminary, auxiliary, and technological time can be calculated using this information, as well as ideas for the equipment and machines required. If the given information cannot be processed to the next phase, the control mechanism must send feedback to reschedule or optimize the process.

The system hierarchy is depicted in Figure 2.3. The CAPP is the system's brain; data is gathered from the current database, and other CAD models are used to create the process plan, operation sequencing, and scheduling. After obtaining the answer from CAPP and evaluating it using the criteria listed in Figure 2.4, product planning software creates

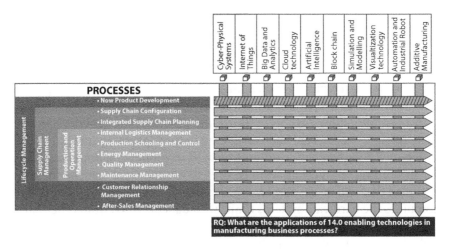

Figure 2.3 Conceptual framework [21] Source: (Zheng *et al.*, 2021).

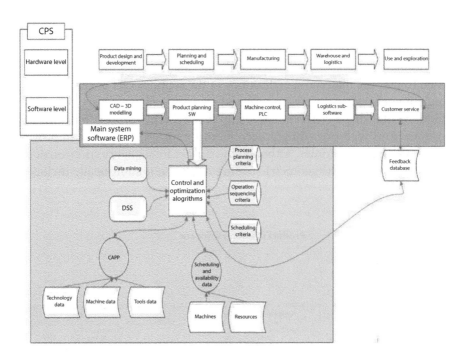

Figure 2.4 Product planning software—schema with connections.

the concludingproduction plan. Cloud-based general system software is connected to product planning software (ERP). It connects the product lifecycle's "product planning" phase to the other stages. The main system receives data from several divisions of the company, starting with product creation and development and ending with user feedback. Data is captured and given to product planning software as information, which may then be translated into knowledge to optimize the phases of the product lifecycle that are highly connected and considered in the I4.0 environment.

Figure 2.5 depicts the criterion tree used in product planning software. The beginning set of criteria will be used to monitor and regulate the process plan that will be developed. For organizational purposes, the software combines the hardware system's limits with data obtained from other parts of the supply chain. The major PP criteria are the product mass and material which is not only the machining features related to the mass, but also related to logistics and material manipulation in the production process [26]. It describes the constraints of vehicles that can be utilized inside the hall, as well as human mobility. Mass limitations can be linked in an I4.0 environment to the limitations of robots and other operators used in the transportation of materials from suppliers or warehouses, as well as the units used in the transportation of finished goods to warehouses, each with its own set of limitations, constraints on product storage [27].

Though material is associated with the machining process, it is also used to manage material specifications in the supplier's set of connections, including checking the ease of use and price as well as delivery time. The fundamental shape criterion speeds up the archiving process and enables comparisons with preceding orders, allowing for uninterrupted improvement with each new order. Although the surface final quality criterion does the same thing as the geometric feature criterion, it additionally includes

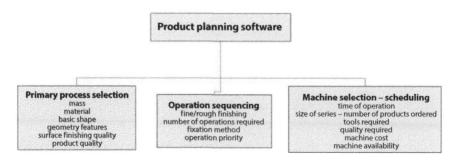

Figure 2.5 PP software—control and optimization criteria [25], Source: (Wollschlaeger *et al.*, 2017).

a cost-benefit analysis. The results of numerous criteria are combined in a product quality criterion, which is then compared to the archive at present with client feedback to identify the optimum product quality and cost.

The second set of criteria is the operation sequencing criteria. Many factors manipulate the order in which operations are carried out in the manufacturing process, but the most important ones are managed and regulated by product planning software. The first is the management of fine or coarse finishing operations.

The necessity of completing rough operations before moving on to finishing operations is the deciding issue. This may appear to be a minor technological issue, but when the entire process is automated, it must be closely monitored to avoid wasteful waste. A statistic for comparing and contrasting processes and sequences is the number of operations required. Because the same technology may be used in various procedures, it is critical to link each one to the next criterion, which is the fixation technique. The adoption of a different anchoring mechanism demands two separate actions for component manipulation. Control and optimization make it possible. For an improve performance, as well as increase production, decrease auxiliary times and cut expenses.

The operation priority criterion at the closing stages of the operation sequencing definition determines the ultimate order. One activity may require completion before another due to various geometrical or organizational needs, which is again recognized via information interchange between CAPP and common system software. The final set of requirements is the arrangement criteria.

Aside from the time consideration, product planning software necessitates device selection, which is heavily dependent on the capabilities of the production system. The system directs the product to determine which machine will do the operations. As originally stated the duration of the surgery is the first factor to consider. Time spent on preliminary, auxiliary, and technological activities the calculation yields the overall time spent on the task. The total quantity of products ordered, as well as the ordered product series, is the next need. Because the machines in use are optimized and scheduled just in time, this is referred to as "just in time scheduling." In real time, a choice is made. If a corporation has a large number of machines that can execute the identical task, the product planning system can perform the same role by providing real-time availability information and sending a particular number of items to units that are available. A variety of machines have been chosen as alternatives. Depending on the tools available and the required performance software allows you to do, a variety of things depend on the situation.

2.6   Industry Production System 4.0 (Smart Factory)

Product planning software is in high demand, and it is intended to identify and analyze various data in real time as a linkage between the common system and the CAPP, not only to obtain the results required for physical manufacturing performance but also to createacquaintance for process uninterrupted improvement and optimization. In a nutshell, it compares the data to the criterion's restrictions, but it is also about the data. It is your obligation to assess the significance of the criteria in diverse situations. In practice, there are two approaches to making this program. The first is a predictive analysis based on data mining. With a significant amount of data, it may be able to find performance trends and problems, allowing it to forecast future behavior (methods). The second way is to use a decision support system that assigns varying weights to various criteria and then generates a final conclusion.

In terms of practical application, combining the two—using data mining to process large amounts of data and a decision support system to make different levels of importance of criteria with predictive analysis to recognize when certain criteria should be prioritized over others—would be ideal [25]. Figure 2.6 depicts an I4.0 smart factory.

The key phase in a reconfigurable manufacturing system is digital to physical conversion. Reconfigurable fabrication systems are the most current breakthrough in the creation of a production system. The beginning stage involved fixed manufacturing lines with machinery specialized for specific tasks, enabling the creation of only one product. The next phase was flexible production systems, which used programmable machines to generate a wide range of things but lacked capacity flexibility. According to recent research, reconfigurable manufacturing systems are capable of adjusting their hardware and software components to satisfy ever-changing market needs for product kind and quantity.

In I4.0 factories, CPS, which are physical systems with ICT components, are used. They are self-contained systems capable of making decisions based on Machine Learning algorithms, real-time data collecting, analytics results, and previous victorious behaviors. In general, CNC (computer numerical control machinesand NC) are used, with self-organizing and self-optimizing mobile agents and robots accounting for a significant portion of embedded sensors in smart industrial products collect real-time data for geolocation, product monitoring, and ambient factors across a wireless network. Smart devices can also control and process information. Customers may have their control on logistical path through the state.

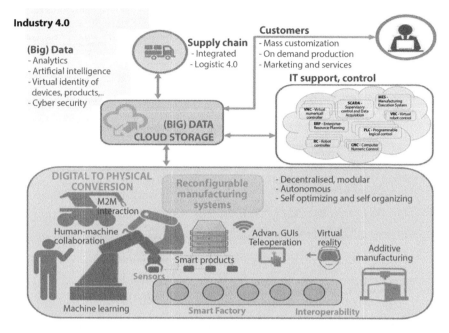

Figure 2.6 Industry 4.0 smart factory [20], Source: (Rojko, 2017).

This permits condition-based maintenance, which is essential for items that are part of larger systems (like for example power converters in electric grids). Production elements in I4.0 have a virtual identity, which is a data item saved in the data cloud, in addition to their physical representation are depicted in Figure 2.7. This virtual identity can store papers, 3-D models, unique IDs, present status data, historical data, and measurement/test data, among other product-related data and information. Interoperability and connection are also important aspects of I4.0. M2M interface, production systems, and actors all demand a steady flow of data amid devices and components. The industrial IoT allows machines, products, and factories to communicate and interact (mostly based on wireless networks). Human-to-machine (H2M) collaboration is another key issue, as some industrial jobs are too unstructured to be entirely automated. Collaborative robotics is generating a lot of buzz in the scientific community right now.

Human workers and specially constructed compliant robots collaborate on the manufacturing production line to execute complex and unstructured work assignments. Previously, such tasks were entirely completed by hand. Advanced user interfaces are being developed for innovative forms

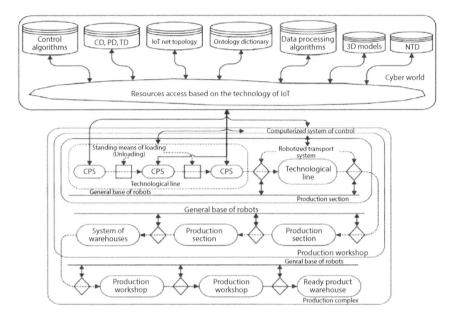

Figure 2.7 I4.0 support and operational level control in the industry [28], Original source: (Zakoldaev *et al.*, 2019).

of M2H communication. They often involve teleoperation and are based on augmented reality surroundings.

Additive manufacturing, such as 3D printing, is frequently indicated as one of the major technologies in I4.0 manufacturing. A straight digital thread from design to production can be created using quick prototype technologies like 3D modeling. Manufacturing facilities can be established, allowing for a faster period from concept to product. Additive manufacturing, on the other hand, has yet to achieve the same level of success. New materials must be created to provide the same level of quality as old industrial methods.

2.6.1 IT Support

In the operation of an I4.0 smart factory, the use of software tools is crucial. The present production system's support software pyramid structure. The ERP tool isutilized at the business level. SCM, sales and distribution, accounting, human resource management, and other facets of corporate planning are all supported by ERP and other enterprise-wide planning. Commercially available solutions are employed the majority of the time.

SAP, a German company, is the most preferred solution at the moment. In typical ERP software, the decision-making process iscentralizedat the top of the automation pyramid. The bulk of ERP solutions does not allow for quick PP adaptation on behalf of unplanned events. Manufacturing execution systems (MES) is the second rung of the classic automation pyramid. Production reporting, scheduling, dispatching, product tracking, maintenance operations, performance analysis, labor tracking, resource allocation, and more are among the other features. Topics, such as shop floor management and enterprise (business) system communication, are covered. The majority of commercial software solutions are centralized and not dispersed to shop floor personnel. This is a crucial limiting issue when flexibility is required due to the volatility of the customer's order flow and/or a changing manufacturing environment, including shop floor configuration. Machine/device level controllers, such as programmable logic controllers (PLCs), robot controllers, and other controllers, are the next operative level, which is based on supervisory control and data acquisition (SCADA) control system architecture. The automation pyramid's final phase is the machine/device level. Unlike the top two layers, this layer has a naturally dispersed control level. ERP and MES technologies have been in use since the 1990s and are critical business software.

Both systems are often modular, but they function in a centralized fashion, restricting the capacity to dynamically change the production plan. Traditional ERP and MES systems, on the other hand, should not be seen as major hurdles to the adoption of the I4.0 concept, but as a step in the right direction. The adoption of a standard MES tool, in particular, already needs contemporary IT infrastructure on the shop floor, which is also a requirement for further progress toward a smart factory. The next big difficulty is the integration of data across ERP, MES, and other software toolsused in the company, such as customer relationship management (CRM) (to help manage connections with the outside world) and business intelligence (for business analysis purposes). Among the challenges that must be addressed are database integration and communication methods. The old automated structure, it may be argued, is not the best solution for I4.0 since it is not stretchy enough to respond to volatile changes in the order flow and on the shop floor. Reconfigurable manufacturing systems are better served by distributed MES solutions, which decentralize the majority of functions. For complete support of reconfigurable systems, a constant flow of information (vertical and horizontal integration) amid all rudiments is essential [29].

2.7   I4.0 in Functional Field

2.7.1   I4.0 Logistics

Demand for extremely customized goods and services are growing. As a result, inbound and outbound logistics must adapt to the shifting landscape. Environment. It can no longer be managed using traditional methods due to its rising complexity. The term "logistics 4.0" refers to the use of logistics in new and innovative ways. In combination with the breakthroughs and uses of cps Logistics 4.0 is a subset of logistics 4.0. The term "smart logistics" is being defined in the same way. The definition of "smart products" was based on a technology-driven methodology. The term "smart services" refers to products and services that are capable of completing tasks. That would ordinarily be done by people. They also allow employees to delegate responsibilities so that they can focus on more important tasks. More intellect than automated methods, or the intelligence of a plain smart product or [30] smart service can provide.

2.7.2   Resource Planning

"Smart logistic" is a logistics system that can help an organization become more flexible, react to market changes, and get closer to its consumers' needs. Better customer service, manufacturing optimization, and lower storage and production costs will all be possible as a result of this. Because "smart logistics" will evolve in response to actual technological breakthroughs, it is vital to identify the technology's current state. The increased use of the internet, which permits real-time communication between machines and humans, as well as greater digitization, has resulted in this new paradigm. A resourceful and powerful logistics 4.0, in our opinion, must rely on and use the following technical applications:

- Resource allocation,
- Systems for warehouse management,
- Shipment planning and management systems,
- Transportation systems that are intelligent, and
- Information protection.

Resource planning management approaches will increase overall efficiency, adaptability, and responsiveness to supply chain changes as a result of the adoption of the I4.0 paradigm and the installation of CPS [30]. Increased visibility and transparency, as well as effective alignment and integration of the supply chain's primary actors, will make sure an adequate forecast of resources (people, materials, and equipment), facilitating resource/process optimization, market alignment time, and asset employment. The required amount of sophistication, as well as the degree of human specialization. As the Internet of Things grows in popularity, so will the demand for resources. As the I4.0 paradigm takes hold, human resource (HR) competencies will alter substantially. It is still being used. The importance of analytical and computational skills, as well as the integration of these talents, is growing. Traditional labor profiles will be altered as a result of technological advancements.

2.7.3   Systems for Warehouse Management

Warehouses have long been important in the transfer of commodities through a supply chain. In today's economic circumstances, however, they must function as a vital source of competitive benefit. Logistics suppliers gain an advantage. The implementation of the I4.0 paradigm will have a significant impact on how warehouses operate today. The creation of "smart" management, particularly through suitable adoption and implementation. Warehouse management systems (WMS) will be installed, which will convert warehouses. Inbound logistical requirements in compliance with industry standards to the new paradigm are 4.0.

2.7.4   Transportation Management Systems

Throughout all value chain phases, the necessary integration between the supply chain's multiple actors and stakeholders will offer total coordination and alignment. Transporters, for example, will be able to report their current location and estimated arrival time to an intelligent warehouse management system, which will be able to choose and set up a suitable shipment. Docking space, enabling just-in-time and just-in-sequence delivery to be optimized. The RFID sensors will reveal what has been delivered while also conveying track-and-trace data across the supply chain. The WMS will assign storage space and request the appropriate equipment to carry the products to a suitable location based on the delivery parameters. Once pallets have been relocated to their allocated locations, tags will send signals to the WMS. Enabling real-time visibility into inventory levels and potentially eliminating costly out-of-stock situations, as well as improving management decision-making skills for any necessary inventory modifications.

A transportation management system (TMS) is a logistics transportation component of supply chain management (SMC). An order management system (OMS) can interface with a distribution center (DC) or a storehouse via a transportation management system (TMS). TMS has been used to help companies control and manage ever-increasing cargo costs, amalgamate with other supply chain technologies (such as WMS and global trade management systems), and handle electronic communications with customers, trade partners, and carriers as technology have advanced. As the scope of their solutions has grown to include these and other features, TMS has become a popular choice for companies of all sizes and across a wide range of industries. Given the expanding use of IoT and the ultimate transition to I4.0, a TMS system is unquestionably a critical component of the logistics 4.0 concept. Logistics 4.0 uses real-time and inline data to increase the efficiency and efficacy of the logistic process. A TMS system enables a company to use GPS technology to track its own vehicles on theroad, supervise freight movement, confer with carriers, combine shipments, and interface with intelligent transportation systems (ITS).

Year after year, TMS capabilities develop, and more companies are expected to employ them in the near future to develop overall transportation management and customer service. As cloud services and cloud computing become more widely available, TMS is becoming more cloud-based. Major software companies are rapidly transferring their TMS solutions to the cloud, reducing the number of on-premise installations significantly in the future. Because the most recent TMS provides greater end-to-end supply chain visibility, they are being adopted by small- to medium-sized organizations, and TMS boosts the upper end of return on investment (ROI) with increased use of mobile devices and services, TMSs are changing corporate strategy.

Smartphone apps will be included in TMS solutions, giving drivers a "breadcrumb" view of where specific trucks are at any given time. In the transportation and logistics industry, IoT and TMS will become increasingly important. Transportation will become more efficient when more physical things are equipped with bar codes, RFID tags, or sensors and logistics firms can track the flow of tangible goods in real time. Objects from a point of origin to a point of destination along the supply chain, including manufacturing, transportation, and distribution are all aspects of the business. Transportation networks and automotive services could be transformed by the Internet of Things. IoT technology can be used as cars gain more advanced sensing, networking, communication, and data processing capabilities. To improve sensing, networking, communication, and data processing capabilities in vehicles. We can, for example, employ IoT technology to track each vehicle's current location and analyze its performance, progress, and forecast where it will be in the future [31]. In the not-too-distant future, a clearly specifiedSmart TMS will be a preset and configured TMS that interacts with IoT devices. Vital for improving the quality of management decision making, and making SCM more flexible and effective, ultimately leading to a comprehensive logistics 4.0 operation.

2.7.5   Transportation Systems with Intelligence

Transportation management, control, infrastructure, operations, policies, and control mechanisms are all interconnected in an Intelligent Transportation System (ITS). It makes use of computer hardware, positioning systems, sensor technologies, telecommunications, data processing, virtual operations, and planning approaches. Because we live in such aglobalized society, the infrastructure that supports it is becoming increasingly crucial. Although the idea of combining virtual technologies with transportation is new to the industry, it is crucial in tackling global concerns. It is critical for improving safety and reliability, as well as travel speeds and traffic flow, and lowering hazards, accident rates, carbon emissions, and air pollution. A reliable transportation platform and cooperative solutions are provided by an intelligent transportation system. Just a few of the applications include electronic toll collection, highway data collection (HDC), traffic management systems (TMS), vehicle data collection (VDC), transit signal priority (TSP), and emergency vehicle preemption (EVP). It is something that can be put to good use.

It can be used in navigation systems, air transport systems, water transport systems, and rail systems, and is not confined to automobile traffic. For system operations and personal contextual mobility solutions, generation 4.0 employs multimodal systems, such as personal mobile devices, automobiles, infrastructure, and information networks. Its cooperative systems technologies play a crucial role in sustaining and growing the industry. The logistics process and fleet effectiveness have improved dramatically, allowing the transportation community to considerably benefit. Improve its results, both monetarily and in terms of long-term viability, it will get better. In the near future, managerial decision making will improve in quality and become more flexible and efficient.

Using real-time and inline data from vehicular ad hoc network (VANET) systems, sensor networks, drone points, and business intelligence, logistics can be improved in the future, thanks to the convergence of machine-to-machine (M2M) communication and cooperative systems technologies systems of intelligence. In a fully operational logistics system, intelligent truck parking and delivery area management; multimodal cargo, i.e., supporting planning and synchronization between different transport modes during various logistic operations; CO_2 footprint estimation and monitoring; priority and speed management can all be done. Environment 4.0 ecodrive support, that is, support for true self-driving vehicles; ecodrive support, that is, support for truly self-driving cars.

2.7.6   Information Security

The development of Internet-based applications has revolutionized how organizations run, thanks to the rise of cloud-based systems, the internet of things (IoT), Big Data, I4.0, BYOD (bring your own device), and choose your own device (CYOD) trends. Organizations are eager to learn about new technology initiatives that have low running costs so that they may provide better and more innovative services and acquire a competitive advantage. However, as companies become more reliant on technology to acquire a competitive edge. One of the most important and difficult pre-requisites for running a successful organization is information security. New technology solutions, in reality, are replete with defects that regularly reveal unanticipated security threats.

In this context, businesses must take steps to safeguard the security of their information assets and IT infrastructure. The increasing amount of data that organizations must deal with on a daily basis, as well as the increasing number of employees, are all issues that must be taken into account. On the one hand, the rise of online transactions, the integration of new systems, the greater theoretical possibility of third-party access, as well as a lack of computer security awareness, all serve as larger motives to exploit not just software but also human weaknesses. Users are generally willing to accept new technologies despite their inherent security weaknesses provided the benefits exceed the risks. Fostering and promoting a security culture, as well as acknowledging that all technological applications and systems have inherent vulnerabilities and that people are still, and always will be, the weakest link, will undoubtedly help businesses achieve adequate levels of security and thus get closer to their business objectives. Monitoring and early detection are also key parts of security management because they allow businesses to respond faster to situations that are more difficult to recognize and explain.

Businesses are increasingly relying on quick responses to security incidents and the installation of preventive security measures as a competitive strategy. To offer an appropriate level of security and ensure business continuity, businesses should plan, implement, monitor, andanalyzethe most effective set of controls. In most cases, businesses identify and pick security.

Procedures that are tailored to their company's demands and security needs. This security requirement should be expressed explicitly in the information security policy, which will also specify the controls that will provide the necessary protection. The next stage should be to analyze and certify the implemented controls to confirm that they meet the information security standards.

In reality, the number of businesses requesting security certification has increased. Financial institutions and insurance firms are increasing their demand for security audits, and businesses can employ certification to meet that demand. It also boosts confidence in a company's capacity to put in place proper security policies for handling and securing secret and sensitive client and corporate data. It is also vital to track and analyze the controls in place to see if they are performing as they should, i.e., measures are in place, are well designed, are operating efficiently, and are evaluated on a regular basis, in order to limit risk exposure.

The International Organization for Standardization/International Electrotechnical Commission Joint Technical Committee (ISO/IEC JTC) has published the ISO/IEC 27000 family of standards, while the National Institute of Standards and Technology (NIST) has a number of publications, particularly the 800-series, which cover a wide range of computer security issues. In the realm of security information, these standards are a significant resource. They also share many security ideas and rules, so adopting those helps firms demonstrate their commitment to protecting information assets and maintaining the confidentiality and integrity of customer data. They also provide their clients and partners more trust in their ability to prevent and quickly recover from production or service interruptions [29].

2.8   Existing Technology in I4.0

2.8.1   Applications of I4.0 in Existing Industries

I4.0 helps to improve workplace conditions and safety by utilizing smart machines. Manufacturing rates are boosted by smart technology, better robots, and cloud computing. These technologies enable higher quality items to be produced at a lower cost [9]. IoT connects machines to the internet so that data may be analyzed and supply chains can be digitized. It contributes to enhancing the productivity of the industrial system [25, 32]. In order to perform crucial activities in real time, I4.0 relies on a cloud platform to store and retrieve data. It increases data exchange throughout the manufacturing system and analyzes data from the customer's perspective, as well as market trends and correlations.

2.8.2   Additive Manufacturing (AM)

In I4.0, AM is critical. It converts three-dimensional CAD data or 3D scanner output into a three-dimensional physical object. It builds a solid object by heaping components on top of one another. It is made from metal, thermoplastics, ceramics, biochemicals, and other materials. Stereolithography (SLA), selective laser (SLS), fused deposition modeling (FDM), direct metal laser sintering (DMLS), polyjet 3D printing (PIP), inkjet 3D printing (IJP), laminated object manufacturing (LOM), color jet printing (CJP), electron beam melting (EBM), and multijet printing are some of the technologies used in am (MJP).

It creates a customized product that meets the customer's needs in less time and for a lower price. This technology can quickly introduce a new product because to its design and production flexibility. There are far less wastes because this method creates products layer by layer. It can be used for a variety of medical purposes, including the creation of patient-specific implants, equipment, and gadgets that are suited to their unique requirements [6, 33, 34].

2.8.3   Intelligent Machines

A smart machine in I4.0 can greatly reduce human labor by leveraging automated/intelligent processes. Artificial intelligence (AI) and deep learning/machine learning are two cognitive computing technologies that are embedded in smart machines. These technologies are also used to help people with problems like making decisions and taking action. In the future, these technologies will be able to deliver efficient production techniques.

A smart machine will take over a human's job and be useful in medicine by recommending the best treatment options and accurately diagnosing diseases [14, 35].

2.8.4   Robots that are Self-Aware

An autonomous robot is a key component of I4.0 since it can complete tasks without the need for human contact. These can be used to collect information about the environment. The usage of robots in industry has been shown to increase the efficiency of production systems. These are used, among other things, to increase system performance, material handling, movement, and control. If a person feels threatened at work, he or she can use it [36, 37].

2.8.5   Materials that are Smart

In the fourth revolution, smart materials are also used to improve product performance. These materials have the ability to change shape over time or as a result of temperature changes. A smart product may alter shape to meet the demands of the user and can be used in a variety of medical settings [38, 39].

2.8.6   IoT

Sensors, actuators, and physical devices are used to aggregate data for decision-making processes using network facilities. The Internet of Things improves a product's value and utility [40]. It incorporates the physical world into a computer-based system, reducing human intervention while increasing economic value, efficiency, and accuracy [41–44].

2.8.7   The Internet of Things in Industry (IIoT)

The IoT is a subset of the internet of things that focuses on two key areas of the industry. It aims to boost efficiency while simultaneously improving safety and health. In I4.0, the IIoT uses big data and analytics to integrate a CPS and manufacturing process. It regularly examines industrial systems and practices to maintain openness [45–47].

2.8.8   Sensors that are Smart

A smart sensor is a device that can collect data from the physical world and analyze it using computer resources. Temperature, humidity, location, pressure, leak detection, sensor, accelerometers, and other processes are all monitored and controlled by sensors in I4.0. Depending on the needs of the sector, several types of sensors are used in various real-time data applications [48, 49].

2.8.9   System Using a Smart Programmable Logic Controller (PLC)

A smart plc is a programmable device that controls events like lift movement, pump operation, turbine rotation, motor on/off, and warnings, among other things. It is the hub of an industry that deals with a wide range of industrial operations. It is utilized to keep track of and operate entire industrial processes like assembly lines, robotic devices, and other activities that demand ongoing attention [37, 49].

2.8.10   Software

Traditional manufacturing and product development processes must be radically overhauled. Before parts are built, the software is also utilized for rapid design, analysis, testing, research, and experimental investigation using various methodologies and theories [39, 50].

2.8.11   Augmented Reality (AR)/Virtual Reality (VR)

These are used to provide real-time information by integrating with humans and electrical systems more effectively. These tools aid in the detection of any flaws or failures. These contribute to enhancing planning quality, efficiency, and worker safety by having a shorter development cycle.

These collect sensory inputs such as video, pictures, music, haptics, and GPS data from computer-mediated reality concepts and connect them to the physical/real environment [51, 52].

2.8.12   Gateway for the Internet of Things

In I4.0, IoT gateways are used to collect, modify, and store data from smartphones, tablets, and other similar devices. It facilitates data analysis and networking. IoT gateways are also utilized to facilitate communication between on-site processing, the cloud, the field, and other storage options [53, 54].

2.8.13   Cloud

For quickly upgrading and extending business sectors, cloud computing is necessary. Remote services, management, performance, and benchmarking are just a few of the applications available. The cloud is also utilized to expand technological resources and improve the flexibility of reprovisioning [5, 50].

2.8.14   Applications of Additive Manufacturing in I4.0

An open platform communication (OPC) server is a server that links all legacy devices to a cloud or gateway for secure data storage and analysis.

It can also link objects and machines, as well as IoT gateways and other devices [55, 56].

2.8.15   Artificial Intelligence (AI)

This includes all aspects of learning, thinking, and self-correction. Machines and artificial intelligence (AI) systems are created to imitate human intelligence. It is also a big aspect of I4.0, which aspires to make machines act like humans. It comprises the use of a computerized program to the automation of specialized machinery in order to undertake nonrepetitive and difficult tasks with robots efficiently. As a result, there is not much that can be done to save money on labor. I4.0 is a connected ecosystem that makes collaboration, monitoring, and automation easier [15, 57–59].

2.9   Applications in Current Industries

2.9.1   I4.0 in Logistics

Bottega Veneta, a worldwide luxury goods firm based in Italy, is implementing I4.0 technology in its manufacturing process, particularly in its logistics. The structural prototype presents a sole-data model that is applied to gather and represent huge volumes of data by all participants in the manufacturing processrelated to the decision support system (DSS) production process so that production can be planned and different scenarios can be focused on, resulting in better decisions. Separate maximization techniques, such as mixed integer programming (MIP) and constraint programming (CP) can be used to model production planning; as a consequence, they have produced a design of this MIP tool, furthermore, they wish to apply CP technology in the upcoming destiny [9].

Each task on a production order necessitates two major decisions: who will complete it and when. Several complex constraints must be addressed to make these judgments, some of which represent technological limitations and others which represent supply chain issues. Finally, several goals must be met, such as reducing late delivery and maximizing available resources [11]. Another option is the DSS tools, which uses the Bottega Veneta logistics, which comprises foreign stocks, on-site production materials, and regional contractors, to assist users in making better activity optimization decisions [60]. The application offers a variety of perspectives on the data that was analyzed as well as the answers that were presented in the present situation. The DSS comprises a diagrammatic picturization of each job order, with a focus on the intricate connection among the many functions that are interdependent, utility survey and project record with a focus on source utilization, and periodic compliance to constrain reports, among other things.

The research concluded in a supply chain information and decision-making system chain. The first step was to gather information. The first stage was to conduct a detailed analysis of manufacturing and production regulations in nearby coordination with process engineers participating in the decision-making process. The ISA 95 standard was used in the second stage to formalize such knowledge structures. Finally, a numerical model was created in the third stage to resolve the trouble across each supply chain actor. Suggested innovation involves the creation of amethodology for collecting and encryptingshapeless knowledge, in addition to numerical formulations of the original model (Martin Holubcik).

2.9.2   I4.0 in Manufacturing Operation

Lamborghini, the world's most famous exotic carmaker, has opted to change gears andstarting from the earliest stage with its new urus, the world's first fabulous sport serve vehicle. The renowned carmaker sought guidance and participation from KPMG's expertise as it set out to construct what would be its first futuristic factory. With a specific consumer market in mind, the march toward I4.0 began. The user and KPMG cooperate on an I4.0 scheme that specified how the production line, its innovation, and its operation working process, in view of a profound comprehension of the needs and inclinations of the designated market bunch. KPMG made and introduced a confounded stage that oversees producing processes all through the whole gathering and complete platform, based on a comprehensive I4.0 plan and its set objectives. The factory has a one-of-a-kind modular design that combines robots and digital sensors in a collective setup.

By combining M2M and robotics communication to accompany Lamborghini's expert shop floor workers, the future facility blends augmented reality and realitymanufacturing, it boasts the most modern I4.0 standards. Automatic-guided vehicles (AGVs) carry each vehicle throughout the manufacturing floor, automatically delivering each vehicle to the appropriate work island.

Workers can use tablets to commandeach step of production at the field level or far off from anywhere, eliminating the need for paper papers. On-site workers have complete control over the production process, thanks to electronic monitoring, data collecting, and reporting that is available quickly from every area of the shop floor or from anywhere in the world. The ground-breaking Lamborghini project, according to KPMG's project leader Carmelo Mariano Italy-based person, underlined the importance of prioritizing value over technology on the I4.0 journey (KPMG).

Bottega Veneta, an Italian high-end goods company known for its leather goods, was the focus of an I4.0 enabling technology application. In particular, incorporating the proposed framework integrates the manufacturing process into the supply chain by introducing a common data pattern that will be utilized by all players in the production method to gather and interpret the huge quantity of data created. A decision support system (DSS) permits the production schedule to concentrate onseveral situations to make better judgments. The suggested solution is ADSS that would guide users to makesuperior operational choices throughout Bottega Veneta's complicated distribution chain, which in corporate outwardfeedstock suppliers, internal manufacturing job shops, and regional sub apprentices [22]. The program provides several viewpoints on the data that has been analyzed, as well as the solutions that have been proposed in the situations that have been selected.

The DSS gives a graphical portrayal of every creation request, zeroing in on the perplexing connections between the different undertakings needed by each request; reports zeroed in on every one of the assets that were the subject of examination and arranging, with a specific spotlight on their utilization; records that give knowledge into the arrangement's ability to meet or not meet the needed deadline. An unmistakable production network part might be utilized for every activity in this model work routine for the assembling of a completed thing [46].

For obvious industrial confidentiality reasons, the subtleties in each crate have been intentionally anonymized, yet the actual portrayal is a legitimate exhibition of how the DSS might furnish the client with an engineered at this point total depiction of a genuine creation request.

In actuality, Figure 2.8 is a self-generated output from the DSS for every investigated and analyzed case. The job schedule that has been scheduled A significant amount of data is encoded using this method. Graphical elements that presentthe client with a right away reasonable insight about each request, would commonly necessitate dexterous manoeuvring through hundreds of spreadsheet columns [61].

- Tasks that have been done earlier are represented by grey text inside the boxes, while errands that had been arranged utilizing the DSS are represented by black text inside the boxes.

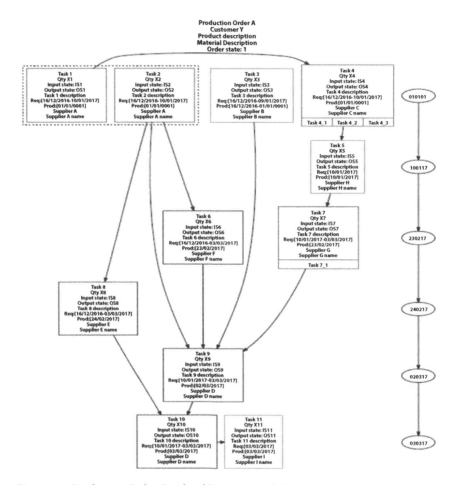

Figure 2.8 Production Order Graphical Presentation [20], Source: (Rojko, 2017).

- The arrows indicate the order in which tasks must be completed.
- Technology constraints force the jobs to be processed at the same time, as seen by the dotted lines boxes made up of other boxes.
- The type of supply chain component in charge of job execution is indicated by the color of each box's frame: green farmed boxes are processed [20].

2.10   Future Scope of Research

2.10.1   Theoretical Framework of I4.0

I4.0 technologies can be categorized into at least two levels based on their principal aim, according to our conceptual framework. The framework's "front-end technologies," which consider the transformation of industrial operations based on new technologies (smart manufacturing) and how goods are delivered, are at the forefront (smart products). It also covers the distribution of raw materials and completed items (smart supply chain), as well as new ways for employees to execute jobs using contemporary technology (smart working) [22]. Because the four "smart" dimensions are concerned with operational and market needs, frontend technologies are suited for this technological layer [44]. As a result, as represented by the schematic arrow in Figure 2.9, they play an end application role in the value chain of the enterprises. It is worth emphasizing that smart manufacturing is the frontend technology layer's central dimension, with the other dimensions tied to it.

A second layer, named "base technologies," supports the frontend layer and contains technologies that provide connectivity and intelligence for frontend technologies, as shown in Figure 2.9 [21]. This final layer is what distinguishes the I4.0 idea from the preceding stages of industrial development. This is the case because base technologies allow frontend technologies to be connected in a fully integrated manufacturing system. The subsections that follow define each of the layers in us Figure 2.9 structure. The purpose is to determine how these technologies are deployed in manufacturing firms and whether any trends emerge [22].

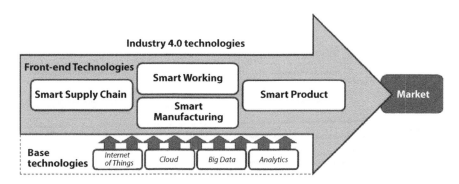

Figure 2.9 I4.0 technologies theoretical framework.

The overall picture of the adoption of I4.0 technology states that the structure was divided based on the conceptual framework of Figure 2.10, which was expanded based on empirical findings. These intensities add to the complexity of implementing a set of actions. The key difference between these models and the framework is that the framework suggested ideal stages, whereas the industrial sector is grounded in empirical data. The focus of the detailed technologies was mostly on I4.0 capabilities. This model is also more comprehensive, as it takes into account not only internal smart manufacturing technology, as other models do, but also a host of other crucial dimensions and technologies.

This framework will guide the subsequent discussions. Several earlier hypotheses were confirmed as a result of the research. One of them is the industry's level of implementation. The 4.0 concept is influenced by several elements, including the scale of the organization, the relationship between large businesses, and the sophistication with which I4.0 is implemented [54].

This is consistent with general innovation theory, which states that large companies are more likely to invest in process and product innovation because it necessitates large investments in technology infrastructure, which small enterprises cannot afford. Furthermore, advanced adopters, not just a few, are at the forefront of all advances. Specific, which could mean that the evolution of I4.0 technologies involves putting together technical solutions like "lego" rather than swapping one for another. The

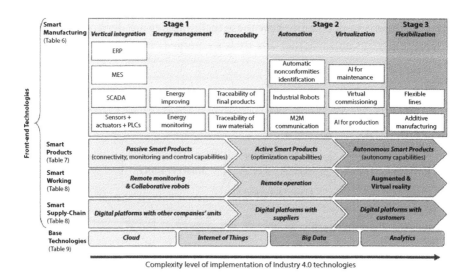

Figure 2.10 Framework for summarizing the findings of I4.0 adoption patterns.

increasing introduction of technology in the maturing I4.0 has resulted in a surprising increase in flexible line acceptability.

One of the I4.0 concepts that have been offered is the flexible line, which can be enabled by the use of additive manufacturing to produce a variety of components and products on the same line [55]. Making a present manufacturing line more flexible will necessitate adjustments to the layout and production procedures in addition to the adoption of new technology. This could be prohibitively expensive, or it could necessitate an excessive number of people. Adjustments are being made, which is causing operations to be affected. As a result, more research into the role of women in the workplace is being conducted. In the future, I4.0 will necessitate flexible lines [16].

Business-to-business (B2B) operations, as opposed to business-to-consumer (B2C) models, are largely used in the machinery and equipment industry. For B2B transactions, more specialized and personalized solutions are necessary, resulting in more communication and relationship between the company and the customer. As a result, the relative importance of various front-end technologies evolves. Companies that have progressed their I4.0 implementation, rather than a select few, use the majority of front-end technology. A sequence of implementation phases for the technology used could be drawn. We organized this information into a framework, which is the most significant contribution of our results, illustrating how I4.0 technologies are deployed and linked [59].

2.11   Discussion and Implications

While the company is still in existence, we may gain full details only if we look for the past 30 years, the technologies that are introduced by I4.0 show their potential. I4.0 firms that shine out in each category, according to Matthew Wopata, are:

2.11.1   Hosting: Microsoft

The major supporting partner in the field of I4.0 is Microsoft. It has helped many industrial automation companies. For decades, Microsoft has worked with both end-to-end users (manufacturers) and suppliers (Honeywell, Panasonic, Saxo, etc.) toenhance and function the difficult mission on-premise SCADA and Military Engineer Services. The companies like Emerson, Siemens, etc. are receiving services from Microsoft.

2.11.2 Platform for the Internet of Things (IoT): Microsoft, GE, PTC, and Siemens

There are many varieties of shapes and sizes for IoT (for more information, check IoT platforms study); In the market, there are two types of categories in the IoT platform that are:In the platform of industry IoT (IaaS + PaaS platforms) that are intimately integrated with infrastructure. The IoT solution is strongly connected with the IaaS backend of a cloud provider. About Industrial IoT services, Microsoft has also arisen with leadership among Amazon, IBM, and Google giving IaaS + PaaS platform offerings for industrial firms.

IoT platforms have an agnostic infrastructure. The elements of the IoT stages are arranged in a sequence included in these IoT platforms, although the IaaS/cloud backend layer is purposely vendor-agnostic. The leading sectors in industrial automation businesses are Siemens, GE, and PTC. which allows for more flexibility in multicloud and on-premise hosting.

2.11.3 A Systematic Computational Analysis

The most-attracted industrial analytics start-up is situated in the US, thanks to a $117 million investment in November 2017 that valued the company at $2.3 billion, Uptake started 2018 on a positive note. Uptake purchased Asset Performance Technologies, a SaaS firm that gives the information of nonsuccess for over 800 industries asset categories, in April 2018. Before the end of 2018, it has to accelerate the growth and solutions for predictive maintenance hoped to be uptake, such hopes had not materialized. Based on LinkedIn, Uptake's total employee count had declined to 550 in December 2018 from an all-time high of 800 in January 2018 (year-on-year decrease of 25%). To add insult to injury, in December 2018, GE filed a lawsuit against uptake, alleging that the business committed a ruthless scheme to poach GE officials would eventually expose GE's classified information expertise. Regardless of its new setbacks, absorb garnered a prestigious clientele (including Rolls-Royce and Caterpillar). Former GE executives are leading a seasoned (if not calloused) leadership team that aims to turn the company nearly in 2019 and keep away from repeating GE's mistakes.

Nvidia had established itself as a pioneer in AI education software and hardware, as well as chip manufacturers. They also invest in the CUDA software library for deep neural networks. This boosts the efficiency of the AI software library for Nvidia GPUs, which is especially beneficial for businesses selling robots, drones, and networking machine vision. The requirements for the required hardware for machine leading software used in factories, such as CUDA-capable NVIDIA GPUs, are one illustration of Nvidia's market presence.

2.11.4   Festo Proximity Sensor

Festo is a discrete industry that serves customers worldwide a German manufacturer thatproduces pneumatic, electrical, and motor control technology. Because of its dynamic cooperation in an assortment of I4.0 working gatherings and attention on creating correspondence standards (OPC-UA, RAMI, Automation ML, and others) that ensure vendor compatibility, Festo has become one of the top I4.0 companies. SeveralFesto devices incorporate the CODESYS period control framework, which gives direct interfacing with an assortment of modern IoT stages as per the VDMA standards.

2.11.5   Connectivity Hardware: HMS

HMS has grown into one of the leading I4.0firms focusing on the industry IoT connection. since its acquisition of eWON in 2016. eWON's product becomes the solution for the industrial problems, then began to control engineers and OEMs to oversee andplan PLCs. eWON Flexy 205's unique mixture of remote connectivity, the powerful, and versatile tool used for various tasks such as commercial agreement changes, and industrial IoT multiside platforms such as far off PLC program construction and isolate condition checking through outsider modern IoT multiside stages like GE Predix, Microsoft Azure, and Siemens Mind Sphere.

2.11.6   IT Security: Claroty

Claroty's strategic investors include Rockwell Automation, Schneider Electric, and Siemens. These high-profile investors pay attention to Claroty though in the past 5 years a slew of commercial cybersecurity setups has emerged. They have used clear strategic investors, who should boost, assist, and build Claroty business. Claroty's portfolio detects the rising cost of danger in industrial cyberattacks asmodern resources are associated with IP-based organizations.

2.11.7   Accenture Is a Systems Integrator

The word "Industry X.0." is coined by Accenture. As well as a leader of digital transformation has positioned themselves as a major I4.0 methodology and execution of a consulting company. Accenture has several key industrial automation manufacturers. It has a strong base in IoT.

2.11.8 Additive Manufacturing: General Electric

The industrial additive manufacturing firm "GE Additive" subsidiary has made various remarkable acquirements and speculations in Arcam and Concept Laser are two notable acquisitions (iconic suppliers of electron-beam melting technologies). A marketplace where the contract manufacturers are connected with the clients who require small quantities of manufactured parts like 3D printers. GE has invested in Xometry.

2.11.9 Augmented and Virtual Reality: Upskill

Cisco and Accenture are the sponsors of Upskill. The customer of upskills are Boeing, GE, Jabil, and Toyota. Upskill is a startup. It is used to assist manufactures to improve their operations. According to Upskill, companies see a 32 percent increase in worker performance after installing their Skylight platform, and it is well placed to employ AR/VR to support manufacturing enhance their functions.

2.11.10 ABB Collaborative Robots

ABB is a major manufacturer of automation and robotics. Yumi's collaborative robots, which include single and dual-arm models, are less expensive and adaptable to ABB's standard commercial robots. In the discrete industry, ABB has a large presence in the robotics market and as an automation supplier. It is also well positioned to benefit from the rising need for collaborative robots as creators seek more adaptable and cost-effective methods to automate manufacturing procedures.

2.11.11 Connected Vision System: Cognex

The product line of the Cognex, which was founded in the year 1981, and the range of the vision hardware and software solutions are widely considered as the market icon in themachine of vision arrangements, covers a comprehensive. Vision ProViDi from Cognex is a knowledge engineering-based image analysis software solution aimed at factories. Greater the quality of products and more flexible fabrication, as manufacture. Machine vision systems will become increasingly important, and Cognex is well positioned to capitalize on this trend and incorporate companies, like

Sense Time, Megvii, and ThyssenKrupp, expanding its already strong customer base.

2.11.12 Drones/UAVs: PINC

PINC uses drones, and the AIR (Aerial Inventory Robots) system fuse drones and barcode scanning technologies to furnish businesses to grow more effectively and efficiently. It is also a significant supplier of building and stock control software, which helps to automate the time-consuming procedure of inventory inspections.

2.11.13 Self-Driving in Vehicles: Clear Path Robotics

Clearpath Robotics was a Canadian firm thatenables self-driving car technology and services. In the year 2009, it was established. OTTO Motors has produced a variety of self-driving automobiles by Clearpath Robotics specifically for the transportation of inventories at industrial plants. Manufacturers can move away from inflexible inventory transport methods to more adaptive self-driving functions, and Clearpath Robotics (through its OTTO Motors subsidiary) is promising to profit from the trend. I4.0 is extraordinary and more simplified technology. We must also consider the impact on society and workers. If we keep on collaborating with man-machine or robots or cobots, which may lead to job losses because of ongoing automation—how to address these major changes. Despite this, the concept of industry 5.0 will exist as a hope by many people that the human touch is undervalued.

2.12 Conclusion

Industry 4.0 is a concept that every new business should embrace from the beginning. Complete system automation and the use of sophisticated technologies, such as the Internet of things and big data analysis, reduces the human factor in the process while also changing the professions inside the organization. The process planner is one of the occupations that will be altered. Because every area of the supply chain and manufacturing process is being automated, process planning must also be automated and linked to other parts of the supply chain and manufacturing process. The general notion of the modification is presented in the initial step of study. It necessitates the amalgamation of process planning, operation sequencing, and scheduling into a single entity. This is the "product planning" phase, with the same-named software serving as a link, control, and optimization mechanism between the CAPP with appropriate databases and the ERP system, the company's general software. It produces a process plan with the exact order of operations automatically, then determines where the product will be manufactured among the machines and tools available. The third avenue to be investigated is CAPP itself, a notion that has been known for decades, but it is now time to put it to the test in an Industry 4.0 context using algorithms that are currently accessible and software that can perform calculations faster than previously. The fourth option is to investigate the relationship between data mining and decision support systems in order to improve the efficiency of product planning software by defining data to be collected from different areas of the supply chain and phases in the product's lifecycle [32].

Finally, Industry 4.0 has strengthened the integration of horizontal and vertical value chains. Increasingly prominent personalized market demand compels enterprises to pay more attention to extended, customer-centric product-value chains, including the customer in the product-value chain will expand Industry 4.0's current capacity, allowing customised products to be produced as efficiently as mass-produced ones. Looking ahead to Industry 5.0, additional restructuring of production and product development will redefine manufacturing processes, as well as product definition and value to customers. It is thus anticipated that new production paradigms and scheduling challenges thereof will soon develop and become a topic for future research [62].

References

1. Gilchrist, A., *Industry 4.0: The industrial Internet of Things*, New York, Springer, 2016.
2. Bartevyan, L., *Industry 4.0–summary report*. DLG-expert report, pp. 1–8, 5, 2015.
3. Soldatos, J. and Lazaro, O., Introduction to industry 4.0 and the digital shopfloor vision, in: *The Digital Shopfloor: Industrial Automation in the Industry*, vol. 4, pp. 64–66, 2019.
4. Grabowska, S., Business model metallurgical company built on the competitive advantage, in: *25th Anniversary International Conference on Metallurgy and Materials*, 2016.
5. Wasike, J.M. and Njoroge, L., *Opening libraries to cloud computing: A Kenyan perspective*, Library Hi Tech News, vol. 32, no. 3, pp. 21–24, 2015.
6. Cheng, Y., Tao, F., Xu, L., Zhao, D., Advanced manufacturing systems: Supply–demand matching of manufacturing resource based on complex networks and Internet of Things. *Enterp. Inf. Syst.*, 12, 7, 780–797, 2018.

7. Jeschke, S., Brecher, C., Meisen, T., Özdemir, D., Eschert, T., Industrial Internet of Things and cyber manufacturing systems, in: *Industrial Internet of Things*, pp. 3–19, Cham, Springer, 2017.

8. Holtgrewe, U., New new technologies: The future and the present of work in information and communication technology. *New Technol. Work Employ.*, 29, 1, 9–24, 2014.

9. Moktadir, M.A., Ali, S.M., Kusi-Sarpong, S., Shaikh, M.A.A., Assessing challenges for implementing Industry 4.0: Implications for process safety and environmental protection. *Process Saf. Environ. Prot.*, 117, 730–741, 2018.

10. Schwab, K., *The fourth industrial revolution*, Switzerland, Currency, 2017.

11. Macaulay, J., Buckalew, L., Chung, G., *Internet of Things in logistics, DHL Trend Research/Cisco Consulting Services*, DHL Trend Research and Cisco Consulting Services, 2015.

12. Crnjac, M., Veža, I., Banduka, N., From concept to the introduction of industry 4.0. *Int. J. Ind. Eng. Manage.*, 8, 1, 21–30, 2017.

13. Koch, V., Kuge, S., Geissbauer, R., Schrauf, S., *Industry 4.0: Opportunities and challenges of the industrial internet*, pp. 5–50, Strategy & PwC, 2014.

14. Ivanov, D., Dolgui, A., Sokolov, B., Werner, F., Ivanova, M., A dynamic model and an algorithm for short-term supply chain scheduling in the smart factory industry 4.0. *Int. J. Prod. Res.*, 54, 2, 386–402, 2016.

15. Lu, Y., Artificial intelligence: A survey on evolution, models, applications and future trends. *J. Manage. Anal.*, 6, 1, 1–29, 2019.

16. Saniuk, S., Saniuk, A., Cagáňová, D., Cyber industry networks as an environment of the industry 4.0 implementation. *Wirel. Netw.*, 27, 3, 1649–1655, 2021.

17. Jazdi, N., Cyber physical systems in the context of industry 4.0, in: *2014 IEEE International Conference on Automation, Quality and Testing, Robotics*, IEEE, 2014.

18. Tay, S.I., Lee, T., Hamid, N., Ahmad, A.N.A., An overview of industry 4.0: Definition, components, and government initiatives. *J. Adv. Res. Dyn. Control Syst.*, 10, 14, 1379–1387, 2018.

19. Heuser, L., Nochta, Z., Trunk, N.-C., *ICT shaping the world: A scientific view*, Wiley-Blackwell, Hoboken, NJ, USA, 2008.

20. Rojko, A., Industry 4.0 concept: Background and overview. *Int. J. Interact. Mob. Technol.*, 11, 5, 77–90, 2017.

21. Zheng, T., Ardolino, M., Bacchetti, A., Perona, M., The applications of industry 4.0 technologies in manufacturing context: A systematic literature review. *Int. J. Prod. Res.*, 59, 6, 1922–1954, 2021.

22. Gajdzik, B., Grabowska, S., Saniuk, S., A theoretical framework for industry 4.0 and its implementation with selected practical schedules. *Energies*, 14, 4, 940, 2021.

23. Trstenjak, M. and Cosic, P., Process planning in Industry 4.0 environment. *Proc. Manuf.*, 11, 1744–1750, 2017.

24. Anderl, R., Industrie 4.0-advanced engineering of smart products and smart production, in: *Proceedings of International Seminar on High Technology*, 2014.

25. Wollschlaeger, M., Sauter, T., Jasperneite, J., The future of industrial communication: Automation networks in the era of the internet of things and industry 4.0. *IEEE Ind. Electron. Mag.*, 11, 1, 17–27, 2017.

26. Carvalho, H. and Cruz-Machado, V., Integrating lean, agile, resilience and green paradigms in supply chain management (LARG_SCM), in: *Supply Chain Management*, pp. 27–48, 2011.

27. Hasan, S.F., Siddique, N., Chakraborty, S., *Intelligent transport systems: 802.11-Based roadside-to-vehicle communications*, Springer, New York, 2013.

28. Zakoldaev, D., Shukalov, A., Zharinov, I., From Industry 3.0 to Industry 4.0: Production modernization and creation of innovative digital companies, in: *IOP Conference Series: Materials Science and Engineering*, IOP Publishing, 2019.

29. Barreto, L., Amaral, A., Pereira, T., Industry 4.0 implications in logistics: An overview. *Proc. Manuf.*, 13, 1245–1252, 2017.

30. Valdeza, A.C., Braunera, P., Schaara, A.K., Holzingerb, A., Zieflea, M., Reducing complexity with simplicity-usability methods for industry 4.0, in: *Proceedings 19th Triennial Congress of the IEA*, 2015.

31. Laberteaux, K. and Hartenstein, H., *VANET: Vehicular Applications and Inter-Networking Technologies*, John Wiley & Sons, Germany, 2009.

32. Grieco, A., Caricato, P., Gianfreda, D., Pesce, M., Rigon, V., Tregnaghi, L., Voglino, A., An industry 4.0 case study in fashion manufacturing. *Proc. Manuf.*, 11, 871–877, 2017.

33. Campbell, I., Bourell, D., Gibson, I., Additive manufacturing: Rapid prototyping comes of age. *Rapid Prototyp. J.*, 18, 4, 255–258, 2012.

34. Niaki, M.K., Nonino, F., Palombi, G., Torabi, S.A., Economic sustainability of additive manufacturing: Contextual factors driving its performance in rapid prototyping. *J. Manuf. Technol. Manage.*, 30, 2, 353–365, 2019.

35. Hozdić, E., Smart factory for industry 4.0: A review. *Int. J. Mod. Manuf. Technol.*, 7, 1, 28–35, 2015.

36. Bahrin, M.A.K., Othman, M.F., Azli, N.H.N., Talib, M.F., Industry 4.0: A review on industrial automation and robotic. *J. Teknol.*, 78, 6–13, 137–143, 2016.

37. Pires, J.N. and Azar, A.S., Advances in robotics for additive/hybrid manufacturing: Robot control, speech interface and path planning. *Ind. Rob.: Int. J.*, 45, 3, 311–327, 2018.

38. Paul, S.C., van Zijl, G.P., Tan, M.J., Gibson, I., A review of 3D concrete printing systems and materials properties: Current status and future research prospects. *Rapid Prototyp. J.*, 24, 4, 784–798, 2018.

39. Haleem, A. and Javaid, M., Additive manufacturing applications in industry 4.0: A review. *J. Ind. Integr. Manage.*, 4, 04, 1930001, 2019.

40. Yang, P. and Xu, L., The Internet of Things (IoT): Informatics methods for IoT-enabled health care. *J. Biomed. Inform.*, 87, 154–156, 2018.

41. Da Xu, L., He, W., Li, S., Internet of things in industries: A survey. *IEEE Trans. Industr. Inform.*, 10, 4, 2233–2243, 2014.

42. Branger, J. and Pang, Z., From automated home to sustainable, healthy and manufacturing home: A new story enabled by the Internet-of-Things and Industry 4.0. *J. Manage. Anal.*, 2, 4, 314–332, 2015.

43. Arnold, C., Kiel, D., Voigt, K.-I., Innovative business models for the industrial Internet of Things. *BHM Berg-und Hüttenmännische Monatshefte*, 162, 9, 371–381, 2017.

44. Luthra, S., Garg, D., Mangla, S.K., Berwal, Y.P.S., Analyzing challenges to Internet of Things (IoT) adoption and diffusion: An Indian context. *Proc. Comput. Sci.*, 125, 733–739, 2018.

45. Chen, H., Applications of cyber-physical system: A literature review. *J. Ind. Integr. Manage.*, 2, 03, 1750012, 2017.

46. Boyes, H., Hallaq, B., Cunningham, J., Watson, T., The industrial Internet of Things (IIoT): An analysis framework. *Comput. Ind.*, 101, 1–12, 2018.

47. Zezulka, F., Marcon, P., Bradac, Z., Arm, J., Benesl, T., Vesely, I., Communication systems for industry 4.0 and the IIoT. *IFAC-PapersOnLine*, 51, 6, 150–155, 2018.

48. Udupa, P. and Yellampalli, S.S., Smart home for elder care using wireless sensor. *Circuit World*, 44, 2, 69–77, 2018.

49. Zheng, P., Sang, Z., Zhong, R.Y., Liu, Y., Liu, C., Mubarok, K., Yu, S., Xu, X., Smart manufacturing systems for industry 4.0: Conceptual framework, scenarios, and future perspectives. *Front. Mech. Eng.*, 13, 2, 137–150, 2018.

50. Jeng, T.-M., Tzeng, S.-C., Tseng, C.-W., Xu, G.-W., Liu, Y.-C., The design and fabrication of a temperature diagnosis system for the intelligent rotating spindle of industry 4.0. *Smart Sci.*, 4, 1, 38–43, 2016.

51. Oyelude, A.A., Virtual reality (VR) and augmented reality (AR) in libraries and museums. *Library Hi Tech News*, 35, 5, 1–4, 2018.

52. Bradley, R. and Newbutt, N., Autism and virtual reality head-mounted displays: A state of the art systematic review. *J. Enabling Technol.*, 12, 3, 101–113, 2018.

53. Ukil, A., Bandyopadhyay, S., Bhattacharyya, A., Pal, A., Bose, T., Lightweight security scheme for IoT applications using CoAP. *Int. J. Pervasive Comput. Commun.*, 10, 4, 372–392, 2014.

54. Tu, M., Lim, M.K., Yang, M.-F., IoT-based production logistics and supply chain system–Part 1: Modeling IoT-based manufacturing supply chain. *Ind. Manage. Data Syst.*, 118, 1, 65–95, 2018.

55. Mosterman, P.J. and Zander, J., Industry 4.0 as a cyber-physical system study. *Software Syst. Model.*, 15, 1, 17–29, 2016.

56. Hofmann, E. and Rüsch, M., Industry 4.0 and the current status as well as future prospects on logistics. *Comput. Ind.*, 89, 23–34, 2017.

57. Upadhyay, A.K. and Khandelwal, K., Applying artificial intelligence: Implications for recruitment. *Strategic HR Rev.*, 17, 5, 255–258, 2018.
58. Hirsch, P.B., Tie me to the mast: artificial intelligence & reputation risk management. *J. Bus. Strategy*, 39, 1, 61–64, 2018.
59. Frank, A.G., Dalenogare, L.S., Ayala, N.F., Industry 4.0 technologies: Implementation patterns in manufacturing companies. *Int. J. Prod. Econ.*, 210, 15–26, 2019.
60. Hermann, M., Pentek, T., Otto, B., Design principles for industrie 4.0 scenarios, in: *2016 49th Hawaii International Conference on System Sciences (HICSS)*, IEEE, 2016.
61. Norton, M., *Introductory concepts in information science*, Information Today, Inc, New Jersey, 2000.
62. Jiang, Z., Yuan, S., Ma, J., Wang, Q., The evolution of production scheduling from Industry 3.0 through Industry 4.0. *Int. J. Prod. Res.*, 60, 11, 3534–3554, 2021.

3

Interaction of Internet of Things and Sensors for Machining

Manivannan Rajendran[1], Kamesh Nagarajan[1], Vaishnavi Vadivelu[2], Harikrishna Kumar Mohankumar[3] and Sathish Kumar Palaniappan[4]*

[1]Department of Mechanical Engineering, Kongu Engineering College, Erode, Tamil Nadu, India
[2]Department of Management Studies, Kongu Engineering College, Erode, Tamil Nadu, India
[3]Department of Mechanical Engineering, Sri Krishna Polytechnic College, Coimbatore, Tamil Nadu, India
[4]Department of Mining Engineering, Indian Institute of Technology Kharagpur, West Bengal, India

Abstract

The major development of internet into a network where the interconnection of items and things generates well-groomed surroundings is termed as Internet of Things (IoT). In recent times, a modern digital manufacturing industries are formed with the help of adapting IoTs from the conventional mode of manufacturing systems. The reformation of manufacturing industries creates the prominent economic opportunity and results in good product accuracy. The industrial IoTs emerged as a developed digital industry with the implementation of data handling technique and then easily can eliminate the burden in this competitive global world in manufacturing industry. Tool condition monitoring is also an important factor in smart manufacturing industries with the help of various sensors involved to continuous monitor and diagnosis. There is a earlier identification of fault systems during machining operations may reduce the downtime and increase the performance and cost savings.

**Corresponding author*: sathishiitkgp@gmail.com

R. Rajasekar, C. Moganapriya, P. Sathish Kumar and M. Harikrishna Kumar (eds.) Integration of Mechanical and Manufacturing Engineering with IoT: A Digital Transformation, (85–114) © 2023 Scrivener Publishing LLC

Keywords: Sensors, Internet of Things, machining, manufacturing, fault systems

3.1 Introduction

In the recent advanced manufacturing industries, the researcher world has faced tough times about the survival today with a good quality products and production [1]. In order to avoid the negative issues to environment, we need to build up strong solutions to meet the best practices in manufacturing activities [2]. In the year 1985, such basic theory of IoTs to develop the connection among human intervention has been established, various machining processes and technological development are those with connect devices, sensors to access the continual monitoring about the status of manufacturing tools while machining process [3]. Even though through this kind of IoT concepts has described about the assessment of various devices based on machining performance, nearly three decades has taken to achieve the best results practices which include wireless networking, web-based computing, artificial intelligence. In most of the machining technologies, now-a-days, there are numerous new materials introduced to face difficult requirements according to industrial needs. Various modeling software, optimization tools, and such kind of continuous monitoring process have been involved to obtain the good machining performance where sensors play a significant role. Through this kind of stated technological methods, it is possible to observe the quality about the surface finish, material removal rate, wear occurrence, temperature source on the work piece, and some other factors involved during machining process have been studied especially machine learning algorthims [4].

The main objective of manufacturing industries is to achieve the mass production, better in quality and target the larger number of customers [5]. Moreover, continuous changes have occurred in the market field, depending on the new technological advancement and newer competitors. Meanwhile, the manufacturing industries has to be survive constantly to the customer surroundings [6]. To rectify the error occurrence in the manufacturing field, emerging new technology and execution software in terms of IoT were established. This IoT leads to enhance the manufacturing environments in terms of mass production, quality, error-free product, less human source utilization, and more efficiency. With this kind of effective integration along with IoT, devices can produce a good environment to the workers, fastness, and more flexible nature of work [7]. New challenges developed in the fields of embedded and communication are those that help to generate the quality infrastructure consisting of numerous devices.

There is an integration of devices with one another to surroundings, which can be modeled in the IoT. In general, IoT consists of three main model, such as internet related, devices related and knowledge oriented [8]. Some of the fundamental IoT has been listed as sensors, actuator devices, internet-related operations, visualization, and interpretation [9]. There are different kinds of technology developed in the IoTs which comprises of RFID, wireless sensors, data storage [6, 10]. Such essential domains can be involved in IoT to enhance the machining performance as personal and home domains [11, 12], enterprise [13], mobile and utility domain [6, 14].

There are lot of changes involved in the industrial surroundings, including the system services, design development when the introduction of IoTs structure [15]. Intelligent sensors and integrated machines have helped to develop new industrial environment and results in good product delivery, good surface finished machining component. Moreover, if the manufacturing units accept the IoTs, highly safety and secured surroundings has been developed [16]. The established IoT in machining environments marked higher level of flexible in nature, earn various kind of possible integration of active applications. Machining performance has also be monitored through the IoT concepts even in advanced manufacturing industries, like as automation and optimization controllable installed devices [17, 18].

Machining operation is vastly spread in all areas of manufacturing industries especially when considering the energy efficiency of machining operation performance normally lesser in range approximately 30% based on various studies involved in research work [19]. The energy efficiency is much more lesser than around 14.8% when milling machining operation taken into account. Therefore the improvement of energy efficiency is much more essential and reduce energy consumption during various machining operation [20]. Energy reduction in a machining operation creates a challenging one to each and every industrials. To rectify this kind of difficulties during machining operation, most of the researchers can concentrate on the energy efficiency crisis, optimization process on the energy and found various models in order to energy reduction [21–24]. Energy monitoring plays a vital role in order to achieve a booming energy management system [25]. With proper installation of devices, proposed model helps to categorize the unwanted energy consumption during machining.

There are numerous scientists identified that monitoring of energy mainly focus on the machine tool operations. During the machining operations, consumption of energy mostly takes place of machine tool, depending on the surface machining and some other feature operations. Such feature operations are notably like cutting process, spindle speed during machining, depth of cut factor, material removal rate and few other operations also involved during various machining, like turning, milling, drilling, etc [26, 27]. In view of improving the energy management during machining operations, machine tools have been monitored in terms of all energy-oriented data [21]. In recent days, especially on the new developed models called IoT tool helps to improve the overall machine performance and high-quality production in a much more effective way. Most of the literatures noted that a highly efficient production and machining performance occurrence based on IoTs proposed models [28]. For instance, energy reduction management during machining depends on the real-time monitoring approach system, radio frequency identification model developed for the proper execution of practical cases during manufacturing operation and result shows best production rate [29–31].

Most of the industrial machining operations were started to install and establish the continuous monitoring by modern sensors and computational intelligence to forecast the performance and evaluation of cutting inserts and machines. Such parameters during machining like quality of the surface finish, accuracy of machining and insert costs are affected severely because of cutting inserts, breakage of tool, flank wear. Those sensors helped to sense the strength and performance of machining operation continuously and easily identify the defects. Thus, various kinds of analysis and studies are taken and rectified. All manufacturing areas are initiated to choose the best system from various classified artificial intelligence methods and those helps to achieve better innovatory components in majority of the manufacturing industries [32].

3.2   Various Sensors Involved in Machining Process

During machining process, the involvement of sensors has become a developed technology in recent times and its one of the major component in future machining and manufacturing industries, resulting in a greater impact especially in continuous monitoring [33]. The two different types of monitoring system in machining operation are introduced as follows:

- Direct method
- Indirect method

Those methods helped to mark the defects easily in any part of the machining operation and easy to access more accuracy in dimension, surface quality, tool wear.

3.2.1 Direct Method Sensors

The direct method sensors highly cost and hard to access the results accurately during machining operation. The offline sensors are of different types as follows [34, 35]:

- Optical sensors
- Electrical resistance sensors
- Displacement sensors
- Acoustic emission sensors
- Ultrasonic sensors
- Laser sensors

3.2.2 Indirect Method Sensors

The other name for indirect method sensor is online method sensors. This kind of sensor is more feasible and reliable to monitor the performance of a machining operation continuously, without any interruption and identifies the inserts wear performance. The online sensors are as follows:

- Cutting force sensors
- Vibration sensors
- Temperature sensors
- Microphones
- Torque-meter sensors
- Current sensors
- Surface roughness sensors, etc.

In general, most of the cases in manufacturing industries, the following four types of sensors are

- Accelerometer
- Acoustic emission sensors
- Current sensors
- Dynamometers

Those types are more oftenly used to continuous monitoring process and predict the accuracy and performance of machining process without any fault occurrence.

3.2.3 Dynamometer

In the manufacturing operation, cutting force plays a prominent factor and explains in detail about the cutting operation. The quality and dimensions of the surface finish will be evaluated easily from the acquired data of pattern during cutting operations [36, 37]. When machining operation takes place in an industries, the inserts performance and dimension accuracy are monitored without any human intervention by cutting force sensors [38]. If the rubbing force is increased during machining operation where the flank wear of the tool could be continuously observed then it leads to reduction in the performance of cutting. There is a rise of cutting force during machining operation because of flank wear, depending on such cutting process parameters like work-piece conditions, insert material, etc.

Figure 3.1 shows the MiniDyn Dynamometer and three component force links, which helps to determine the cutting forces of such microlevel machining operation in manufacturing industries. In general, cutting forces are commonly classified into two important factors, which are follows:

Figure 3.1 MiniDyn Dynamometer and 3 component force link [39].

Static force
Static force is common and standard one which is most preferably involved on determining the cutting force.

Dynamic force
Dynamic force is superimposed fluctuation one, and it may meet the requirements of highly accurate surface finish machined operation [39]. The cutting forces are also denoted in other two common factors [40]:

- Direct measurements
- Indirect measurements

Direct measurements
This kind of measurements is engaged in both magnitude and direction when cutting force is applied during machining, resulting in good accuracy. During machining operation, the tool is mounted on a dynamometer, which observes the changes that occurred with the help of electrical signals relating to the forces applied on it.

Indirect measurements
When compared with the direct measurements, it results in lesser accuracy, even though it is more suitable to the monitoring process during machining operation and easily identify the state of inserts behavior.

3.2.4   Accelerometer

Vibration is one of the important factors during the operation of machining process where the job is easily adjusted/rotated due to high-speed or less-rigid component [41]. Accelerometer is the best-suited sensor to eliminate the vibration occurrence in machining process without any fault surface finish output product and no cause to machinery elements (Figure 3.2). The prediction of roughness finish and changes in the cutting inserts are the main parameters monitored during various machining, like as in turning, milling. In the milling process, vibrations are merely evaluated since there is noncontinuous state occurred in the cutting zone [42, 43]. The placement of sensors during machining operation is more important to predict the accuracy of the component to eliminate unavoidable changes in cutting inserts.

Figure 3.2 Triaxial accelerometer.

In the tool, continuous monitoring system also called as accelerometer sensor has been effectively used to achieve a better results. However, in the state of accessing the insert status and component position placement has faced lot of realistic difficulties when approached in the monitoring system. Such factors are listed as follows [44];

- Spindle speed should be fixed within a limited range to operate.
- There is problem in acquiring the amplitude signals during machining if the sensor distance gets increased from cutting zone.
- The placement of sensor is very close to cutting part which also cause the defect in the collection of signals.
- Once started the machining operation, there is continuous chip formation takes place between the job and inserts which may hits the accelerometer sensor results in fault reading of signals and complete damage to the sensor.

Since the acquired signals from the sensor may not be perfect like signals collected from other kind of sensors namely dynamometer, acoustic emission signal sensors, accelerometer based continuous monitoring system in machining operation have few merits that are too simple and lesser initial cost.

3.2.5   Acoustic Emission Sensor

Acoustic emission sensor is a type of sensor that captures the signals in the form of energy released during the machining operation from various parts of the machinery, such as inserts, job component, etc. It may generates some changes due to friction among the tool and job component, formation of chips, breakage of inserts, change of material properties due to thermal action, etc [33].

Microphone is commonly used to acquire the acoustic emission signals. It is one of the prominent types of AE sensors in the tool continuous monitoring system. Acoustic sensor is a type of wave's generation during machined the parts. Normally, acoustic sensor is classified into two types:

- Transient signals
- Continuous signals

The former signal indicates the insert breakage during machining or damage occurs in the chip formation from the work-piece. The latter signals denotes the cutting zone and wear occurrence on the inserts [45–47]. Figure 3.3 shows the different types of AE signals during the operation of machining. In addition to this information, such general sources are listed of acoustic emission sensor when applied to machining operation [48].

- Changes in cutting area
- Heat up in the region between chip and inserts interface

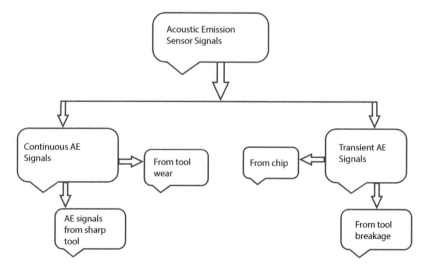

Figure 3.3 Acoustic emission signal during machining process.

- When rubbing takes place between workpiece and inserts
- Smaller size of chips can be scattered and cause damage to workpiece and inserts.

3.2.6   Current Sensors

Armature current is directly proportional to torque, which generates on direct current motor, during rotating, operation of machining is directly proportional to the parameters named as cutting force. As a result, current sensor measurement is used to access the cutting forces during machining indirectly [33]. In this type of sensors, the bandwidth sensible has very limited range based on the moment of rotor movement of motor. Some disadvantages of this sensor are as follows:

- Difficult to acquire the high frequency machinery operations
- High speed surface finish process is not suited since lesser frequency range.

Table 3.1 explains about the respective signal sensors in various machining monitoring process and its applications.

3.3   Other Sensors

In addition to the above-mentioned sensors, few other sensors are also applied to continuous monitoring process during various machining operations, namely temperature, optical and ultrasonic sensors.

3.3.1   Temperature Sensors

In a machining operation, temperature is one of the important parameters in deciding the performance of the tool and product finish. There is a continuous rise or fall in the temperature when exposed to machine the parts between work-piece and tool, depending on the tool wear occurrence changes. Hence, temperature measurement is a best indicator based on insert geometrical dimensions. The cutting temperature also affects the chip formation, surface finish of the product, highly increases the flank wear and even reduces the tool strength. Exact cutting temperature measurement is absolutely hard to determine in each and every machining operations. So that average temperature is taken into account on in and around the cutting zone [49]. Such temperature sensors are practically used as listed below

- Thermocouples
- Thermal resistant elements

Table 3.1 Respective signal sensor in various machining monitoring process and its applications [51].

S. no.	Types of sensors	Initial cost	Applications	Signal accuracy
1	Dynamometer	High	• Easy identification of wear on inserts • Simple in tool breakage findings • Component surface finish is more easier to analysis • Prediction of product accuracy	Highly recommended
2	Accelerometer	Medium	• Surface roughness measurement • Wear prediction	Medium recommended
3	Acoustic emission sensor	Medium	• Insert breakage • Flank wear measurement	Medium recommended
4	Current sensor	Low	• Insert breakage • Flank wear measurement	Recommended

• Semiconductor
• Thermopiles, etc.

3.3.2 Optical Sensors

The optical sensors has been collected on the light beam from surface machined part and then images are reflected with the help of digital system. Then digital information are applied to measure the surface roughness of the machined surface and identifies the status of the inserts. Even-though this optical sensors having few limitations are not suited to inconsiderable surroundings when the usage of various cutting fluid zones like kerosene, lubricating oil, water medium, etc [50].

3.4 Interaction of Sensors During Machining Operation

During machining operation, there are numerous process can be carried out based on sensor developed technologies.

3.4.1 Milling Machining

In a dry milling machining operation, the prediction of flank wear continuously without any physical disturbance where the information has been collected through the data acquisition system and analyzed the performance of the output responses. Figure 3.4 and Table 3.2 shows that the various sensors involved in dry machining CNC milling operation, sensors consists of dynamometer, acoustic sensor, accelerometers. The job is made up of stainless steel material and the three flutes cutting inserts used is tungsten carbide. The parameters can be fixed depending on the need to customer as spindle speed, feed rate, depth of cut and so on. In this operation, dynamometer is enabled to observe the cutting force, piezometer helps to identify the vibration of the inserts, acoustic emission sensor has been utilized to determine the frequency based on the wave formation during machining.

All of the above-mentioned sensors were connected to data acquisition national instruments and extracted from the output responses. The output responses has been collected in the form of signals and segregated in different forms as cutting force, sound signals and vibration. Finally, the acquired raw signals has been analyzed with the help of any IoT-based approach model.

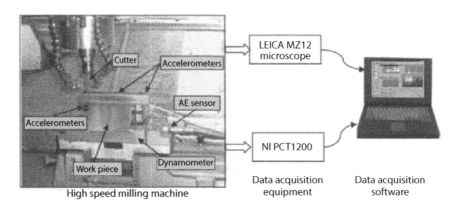

Figure 3.4 Types of sensors involved in milling operation [52].

Table 3.2 Sensors used in milling operation.

S. no.	Machining operation	Sensors	Output responses
1	Milling operation	Dynamometer	Tool failure and flank wear detection
2		Cutting force sensor	Tool breakage
3		Vibration sensor	Sensing of inserts wear
4		Accelerometer	Tool wear monitoring

3.4.2   Turning Machining

Figure 3.5 shows the schematic diagram of turning process with thermocouple sensors. Table 3.3 explains about the different sensors used in turning machining operation. The setup along with thermocouple sensor has helped to determine the cutting insert average temperature. Then, data acquisition card plays a role as a bridge between the sensors and digital Lab-view software in the computer and acquired the temperature information through the Arduino setup. The digital channels of both input and output have been measured and noted continuously for each and every set of conducted experiments. The temperature data has been stored and analyzed in the form of signals which is to be used in the software.

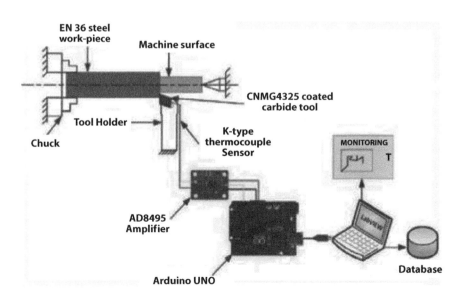

Figure 3.5 Thermocouple sensor used in turning operation [53].

Table 3.3 Different sensors used in turning machining operation.

S. no.	Machining operation	Sensors	Output responses
1.	Turning process	Sensor fusion	Tool failure identification
2.		Cutting force and vibration sensor	Continuous monitoring of insert wear
3.		Dynamometer and piezo electric sensor	Life of tool condition
4.		Acoustic sensor and strain gauge	Tool condition monitoring
5.		Thermocouple sensor	Temperature monitoring of inserts

3.4.3   Drilling Machining Operation

From Figure 3.6, it can be seen that the process of drilling machining and the output responses has been recorded with the help of sensor named as rotating cutting dynamometer, which is fixed on the spindle of motor. In this operational machining setup, workpiece is chosen as aluminum alloy to drill and then drill tool as tungsten-coated carbide.

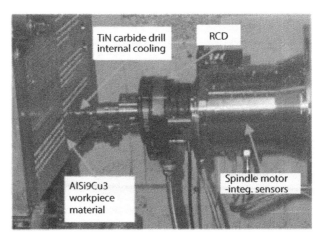

Figure 3.6 Drilling operation with rotating cutting dynamometer [54].

Table 3.4 Sensors in drilling machining.

S. no.	Machining operation	Sensors	Output responses
1.	Drilling	Thrust force	Tool condition monitoring
2.		Acoustic emission sensor	Vibration analysis between drill tool and workpiece
3.		Torque sensor	Tool condition

In addition to this, another sensor is introduced to record the sound waves with the help of acoustic emission, which is clamped on workpiece and then this sensor enables to observe the drilling operation continuously without any interruption until any default detection is identified. Table 3.4 explains about the different types of sensors used in drilling machining.

3.5 Sensor Fusion Technique

It is one of the promising technology in the industries, which works on the basics of human sensory method. Some of the systems like vision, sense of touch, observing noise level, taste of food, etc., operate in a system to identify any kind of object. This kind of method is used to measure flank wear and simultaneously avoid the unnecessary error occurred as like generated in a single way approach sensor. Fusion signal data mean consists of more than one output sensor responses, which can be determined exactly in any kind of machining approach and more effective output performance in minimizing the signal error as much as possible.

There are different of sensors involved in machining operation and those acquired data information can be combined as a single data with the help of neural network, regression-based approaches, etc. The main goal of this approach is to obtain the accurate data after combining of various sensors in tool monitoring. As an example, Figure 3.7 indicates the fusion technology model approach where three sensors are applied, namely dynamometer, surface roughness measurement and accelerometer sensor. As a result from those sensors cutting force, vibration signals and surface roughness values has been collected respectively. Later, all of these recorded signals were combined together through fusion approach with the help of modeling. Thus, tool wear can be predicted from the machining operation where

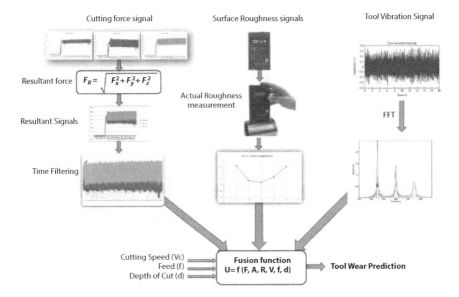

Figure 3.7 Sensor fusion model approach [55].

the cutting process parameters considered as follows are spindle speed, feed rate and depth of cut.

3.6 Interaction of Internet of Things

Industrial IoT has been classified into six various technologies involved and helps to meet the required expectations. These are listed here as follows

- Identification
- Sensing
- Communication
- Computation
- Services
- Semantics

3.6.1 Identification

Identification is the main source that helps to connect every object to the respective address within the stipulated region. It is the one and only method to hint the data information storage. The implementation of this method depends on the client server model and Internet protocol and those that help to allocate the required need of address.

3.6.2 Sensing

Sensing indicates that the collection of information from the sensors and other equipments and then data will be stored in online mode. The industrial IoTs has been described in various formats like smart devices, sensors networks and provide the data at any time based on the needs.

3.6.3 Communication

This is another kind of mode in the I-IoT technologies, which consists of hardware and software data needs to access networking. Most of the essential communication protocols and safety data will be managed with the help of software. Such nearby range software elements used are bluetooth, proximity sensor, radio frequency identification, etc. Wi-Fi, wireless, GSM, etc are those used for wider range. Cable lines, antennas, server wires are the some hardware elements.

3.6.4 Computation

Computation process also expressed in the view of hardware and software like communication process. There is a exchange of data information, system information storage and data analysis can be categorized under cloud based works [50, 56]. There are numerous online cloud based platforms have been emerged in recent times [51]. Hardware can be accessed these kind of process on their required expectations level either through chip control and microcontrollers.

3.6.5 Services

In this way of method, either easier data information collection and hard data information analysis especially in object identification can be differentiated. There are few techniques involved to process the services are big data analysis, machine learning, predictive model based on applications.

3.6.6 Semantics

This is the final stage of process is to brief the data information in and around network. Ontology definition metamodel and model-driven message interoperability standards are the developed model to capture and communicate the ways of data information.

3.7   IoT Technologies in Manufacturing Process

3.7.1   IoT Challenges

Depending on the technology development, each and every technique has some merits to meet out the required expectations, even though the IoTs based operation still has few comfortable challenges. This kind of development mainly indicates that the quality standards and social problems and other essentials are privacy data information and safety. Such concern has been involved in IoT challenges as security, some unauthorized data. Those challenges may interrupt the machining process and result in less production since the need to develop the best automated system and to produce good results in long life services in manufacturing field is based on I-IoTs.

3.7.2   IoT-Based Energy Monitoring System

There are four different layers involved in the energy monitoring which operated based on IoT. They are listed as follows:

- Data acquisition layer
- Information transmission layer
- Data processing layer and
- Application oriented layer

Data Acquisition Method
This is the first foremost method in the IoTs and more answerable one for observing and pre-processing the manufactured collected information and collects the utilization of energy during machining operation in the zone of machine tools and other primary equipments. Moreover, in this segment there are three important elements that play an important role, these are multiple effective sensors, ethernet cards, scanning the codes. The advantages of these are noted as initially the sensors help to determine the exact energy consumed in practice which is mounted on machine tools. Secondly, the ethernet cards are those act collectively along with various process parameters involved in operations like cutting speed, depth of cut and feed rate during the conventional numerical control machining process. And, the final one is as code scanners, it is used to measure the initial

and end of time during each machining process involved on machine inserts and workpiece.

The recorded power consumed data information has been monitored by data pre-processing. It has been maintained in the form hardware and software system as industrial personal computer systems and energy monitoring methods respectively. During machining operation, such parameters has been monitored continuously without human intervention in practical application and those can be recorded if any fault is identified in the state of electric power off, standby time, air cutting and cutting features as shown in Figure 3.8. Then, finally measured the machine operation time and how much amount of energy utilization has been involved.

Data Transmission Layer
The information has been exchanged among the perception and processing layer. Industrial routers and wireless are the main source involved in

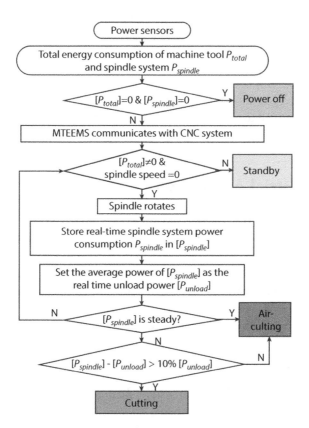

Figure 3.8 Machining operation state [21].

this data transmission. Wireless can be set up on IPCs and then routers attached to server process.

Data Processing Layer

It is the layer helped to predict the performance and energy efficient of the inserts, machining work process and shop floor. In the machine tool process, energy utilization at each and every state of points and total consumption of energy were attained. Based on this parameters, material removal and energy consumption will be clearly analyzed.

In the process of machining operations, numerous jobs have been done at a single stroke of machining, especially machining a number of orders in-line with the process aligned. In terms of energy efficient of machine, quantity, machine process, machine start and end have all to be measured accurately. Now-a-days, advanced system has been implemented to deliver the work process to certain operators with the help of manufacturing execution system. In the machine workshop practice, machine works are to be done inline to energy utilization of the working machine tools and primary equipments.

Application Layer

This is another important layer which is answerable to various kinds of users, like machine shop teams, manufacturer workers and planning team and moreover to all tool operators involved in the process about the energy performance.

3.8   Industrial Application

Figure 3.9 indicates the developed system of IoTs based model to analyze the energy efficient when machining operation takes place. In this proposed model, some features have been established to achieve the requirement needs in this advanced technology world. In this machining operations, energy utilization at various natures of state and climatic conditions could be observed easily whether it lies on the same rate of consumption or any major deviation occurs in the operation.

3.8.1   Integrated Structure

The integrated structure has been established based on two important elements are machining and monitoring system. In each and every element, operation includes hardware and software system. Figure 3.10 shows that

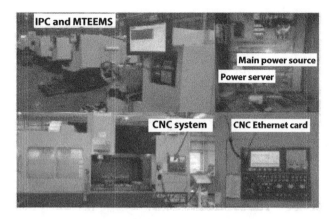

Figure 3.9 Machine shop application.

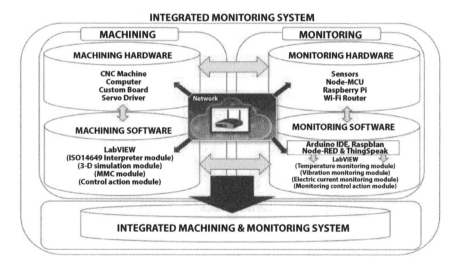

Figure 3.10 Integrated structure of machining and monitoring system [57].

the integrated structure of machining and monitoring where observed the monitoring process of both hardware and software system interconnected with a service based on IoTs. A national instrument Lab-view was used to monitor the continuous status of machining, which is installed in a computer.

Figure 3.11 shows an example of both hardware and software incorporated in a single system. One such illustration has been studied where hardware system during machining includes as conventional numerical computer, custom board, driver system to the server and few sensors are

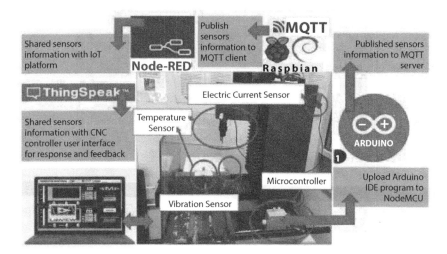

Figure 3.11 Incorporation of hardware and software system [57].

noted as wireless router, MCU and Raspberry model. Interpreted, simulation, machine motion control and control action modules are the four software involved in this machining process. Ardunio, Raspbian, IoT-based program and node red fow are the interfaces used in the above shown machining operations. The monitoring sensors are few included in this operation to discuss are temperature, vibration, current and microcontroller. All of these functions are feed to instruments named as Lab-view.

3.8.2   Monitoring the System Related to Service Based on Internet of Things

Related to IoT-based service system classified into four various kinds of steps as there are perception, communication, application and conventional numerical control milling machine technique. In this machining operation, CNC operation would be observed with the help of monitoring technique devices based on IoT structure in perception layer, and then, data information has been transferred to cloud through the protocol in communication layer and then another stage takes place in application layer where observed data by node red ensures that the interaction among CNC operating machine and things speak sensor and then final stage of process is as the collected data information has displayed in the module fixed system with the help of such platform as Lab-view. Later the output response was confirmed during each and every machining operation of milling where both of hardware and software monitoring module (ex: temperature, vibration and electric current sensors) interacted.

3.9   Decision Making Methods

This is one of the important parameter in the manufacturing industries to decide about the characteristics of quality where the information has been acquired from the sensors. There are various machine learning and deep learning algorithm methods involved in the manufacturing industries. Some are listed as follows:

- Artificial neural network
- Fuzzy inference system
- Adaptive neuro-fuzzy inference system
- Support vector machines
- Decision trees and random forest
- Convolutional neural networks

3.9.1   Artificial Neural Network

This is one of the class of algorithms which is more relevant to the working operation human brain. In general, the artificial neural network architecture are broadly classified as three various layers and named as input, hidden and output layers. The neuron numbers are selected in each and every layers based on the factors are input and output responses, information distribution and presentation to network. The network actually based on the certain structures on layers, activation and function loss. The weighted parameter is involved among the neurons during the time of training model attained and modification of weight has been done depending on the artificial neural network mapping in input to output. This model is mainly used in the prediction of such parameters like surface roughness, tool wear, temperature occurrence between tool and workpiece, etc. Mainly the performance and efficient modeling of ANN has been predicted with the help of back propagation and feed forward neural network approaches. There is better accuracy and good quality based product will be produced. Interpretation of information-based knowledge sharing and objects recognized in each and every node point needs more time, and it can be acted as black box model which is the one demerits involved in this kind of decision method approach.

3.9.2   Fuzzy Inference System

In comparison with the artificial neural network, the number of application oriented operation is less in the fuzzy inference system especially for machining process involved in milling, drilling, turning, etc., if there is known knowledge level enough to understand the process for the model applied using fuzzy inference system. The main merits involved in this fuzzy system over artificial network is that knowledge can be extracted and then executed as if then rules type with the membership as sigmoid, triangular etc. Although the clear instruction is enabled to the model through human brain, it is not able to meet the required level of information. To rectify this kind of errors, a hybrid model development is made by combining the models of artificial neural network and fuzzy inference model, then it is named as neurofuzzy models (Figure 3.12). This model may be eliminates the basic issues that occurred in if then rules by make use of learning method of ANN for automation.

3.9.3   Support Vector Mechanism

The art of splitting the non linear data information is possible in this support vector mechanism and moreover it's a group of machine learning algorithm. In this model, a non linear kind of problems can be derived easily with the help of the function named as kernel trick. The linear form is easily attained from non linear with indirect data and trained. This model can able to solve all kind of high non linear and to determine the learning machine structure. Most probably, this model has been applied in machining operation, like to analyze the tool wear, material removal rate, etc. Table 3.5 explains about the various data-driven methods and algorithms used in this mechanisms.

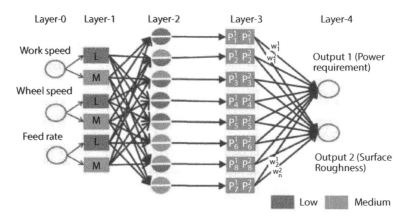

Figure 3.12 Structure of neurofuzzy models [4].

Table 3.5 Various data-driven methods and algorithms used [4].

S. no.	Operations	Objectives of the model approach	Factors	Algorithm used
1.	Grinding	• Prediction of burning • Burning time • Temperature to burn • Surface roughness • Vibration prediction	Grit size, depth of cut, coolant Speed, power, infeed & diameter Grinding force, sensor RMS, time, Speed, feed rate & depth of cut Power spectral density Process parameters Speed, feed rate, depth of cut	Fuzzy ANN ANN ANFIS ANFIS Fuzzy ANFIS
2.	Abrasive water jet machining	• Surface roughness	Transverse speed, water-jet pressure, distance, flow rate and grit size	ANN
3.	Vibratory media finish	• Material removal	Workpiece density, media, time, acceleration	ANN

3.9.4 Decision Trees and Random Forest

With an active base rule if then else is applied to enable the model called decision trees from the output responses. There are some reasons behind the decision trees to handle over hard operations takes place in the machining operations. Moreover, in real practice, progress can be analyzed and actual power can be judged with the help of combining more deep learning model if applying this model helps in improving the accuracy and prediction value responses.

3.9.5 Convolutional Neural Network

This is another kind of model approach, like ANN approach, where network structure is much in depth with in this embedded one. In this approach, it can able to limit the number of factors without negotiating any features and in computational learning algorithm. It can be broadly categorized as many kind layers. There are as follows:

- Convolutional layers,
- Pooling layers,
- Fully connected dense layers.

Based on the chosen of parameters, the resulted responses accuracy is more efficient. Due to many number of layers in this structure and the data representation during trial and error training, it can be grouped together as deep learning. Table 3.6 explains about the merits and demerits of various AI approaches.

Table 3.6 Merits and demerits of various AI approaches.

S. no.	Name of the model approach	Merits	Demerits
1.	Neural Network	The model has been attained without any previous knowledge. Ease to observe at any kind of surroundings with or without complete information data.	There is no exact clarification on designing of neural networks Lack of physical meaning and trail & error method approach.
2.	Neuro-fuzzy model	Combination of both fuzzy systems and neural networks Good extrapolation and capability.	Number of factors need to be study Limited inputs.
3.	Fuzzy logic	It is a simple language and more easier to learn the concepts.	No need to put more effort to learn Limited input responses No regular methods are followed to transfer data from human brain.
4.	Bayesian networks	Presents the casual relationships among variables.	Need high standard data. More digital computational cost is needed to implement the approach.

3.10   Conclusion

In manufacturing industries, there are numerous essential algorithms that analyze about output of machine tools and components where multiple sensing technique are applied to generate more data for the handling purpose of data-driven model approaches. Then, it may help to manage the difficult issues with more accuracy when compared with the analysis of basic numerical methods. There is a very clear indication in the implementation of new digital methodologies to access the efficient output results through the application machine and deep learning approaches. Moreover, the integration of sensors and IoTs to enable the smart manufacturing industries with more accurate continuous monitoring of machining operations and prefect execution of data-driven model approaches.

References

1. Yip, W.S., To, S., Zhou, H., Current status, challenges and opportunities of sustainable ultra-precision manufacturing. *J. Intell. Manuf.*, 1–13, 2021.
2. Rosen, M.A. and Kishawy, H.A., Sustainable manufacturing and design: Concepts, practices and needs. *Sustainability*, 4, 2, 154–174, 2012.
3. Benardos, P.G. and Vosniakos, G.C., Internet of things and industrial applications for precision machining, in: *Solid State Phenomena*, Trans Tech Publ., Kapellweg, Bäch SZ, Switzerland, 2017.
4. Pandiyan, V. *et al.*, Modelling and monitoring of abrasive finishing processes using artificial intelligence techniques: A review. *J. Manuf. Process.*, 57, 114–135, 2020.
5. Colombo, A.W. and Karnouskos, S., Towards the factory of the future: A service-oriented cross-layer infrastructure, in: *ICT Shaping the World: A Scientific View*, vol. 65, p. 81, 2009.
6. Alexakos, C., Anagnostopoulos, C., Kalogeras, A.P., Integrating IoT to manufacturing processes utilizing semantics, in: *2016 IEEE 14th International Conference on Industrial Informatics (INDIN)*, IEEE, 2016.
7. Kalogeras, A.P. *et al.*, Ontology-driven control application design methodology, in: *2007 IEEE Conference on Emerging Technologies and Factory Automation (EFTA 2007)*, IEEE, 2007.
8. Atzori, L., Iera, A., Morabito, G., The Internet of Things: A survey. *Comput. Netw.*, 54, 15, 2787–2805, 2010.
9. Gubbi, J. *et al.*, Internet of Things (IoT): A vision, architectural elements, and future directions. *Future Gener. Comput. Syst.*, 29, 7, 1645–1660, 2013.
10. Biswas, A.R. and Giaffreda, R., IoT and cloud convergence: Opportunities and challenges, in: *2014 IEEE World Forum on Internet of Things (WF-IoT)*, IEEE, 2014.

11. Li, B. and Yu, J., Research and application on the smart home based on component technologies and Internet of Things. *Proc. Eng.*, 15, 2087–2092, 2011.

12. Jie, Y. *et al.*, Smart home system based on IoT technologies, in: *2013 International Conference on Computational and Information Sciences*, IEEE, 2013.

13. Thoma, M. *et al.*, On IoT-services: Survey, classification and enterprise integration, in: *2012 IEEE International Conference on Green Computing and Communications*, IEEE, 2012.

14. Shi, Z. *et al.*, Design and implementation of the mobile Internet of Things based on TD-SCDMA network, in: *2010 IEEE International Conference on Information Theory and Information Security*, IEEE, 2010.

15. Spiess, P. *et al.*, SOA-based integration of the Internet of Things in enterprise services, in: *2009 IEEE International Conference on Web Services*, IEEE, 2009.

16. Zuehlke, D., SmartFactory—Towards a factory-of-things. *Annu. Rev. Control*, 34, 1, 129–138, 2010.

17. Houyou, A.M. *et al.*, Agile manufacturing: General challenges and an IoT@ work perspective, in: *Proceedings of 2012 IEEE 17th International Conference on Emerging Technologies & Factory Automation (ETFA 2012)*, IEEE, 2012.

18. Radziwon, A. *et al.*, The smart factory: Exploring adaptive and flexible manufacturing solutions. *Proc. Eng.*, 69, 1184–1190, 2014.

19. Cai, W. *et al.*, Fine energy consumption allowance of workpieces in the mechanical manufacturing industry. *Energy*, 114, 623–633, 2016.

20. Gutowski, T., Dahmus, J., Thiriez, A., Electrical energy requirements for manufacturing processes, in: *13th CIRP International Conference on Life Cycle Engineering*, Leuven, Belgium, 2006.

21. Chen, X. *et al.*, A framework for energy monitoring of machining workshops based on IoT. *Proc. CIRP*, 72, 1386–1391, 2018.

22. Rajemi, M., Mativenga, P., Aramcharoen, A., Sustainable machining: Selection of optimum turning conditions based on minimum energy considerations. *J. Cleaner Prod.*, 18, 10-11, 1059–1065, 2010.

23. Camposeco-Negrete, C., Optimization of cutting parameters for minimizing energy consumption in turning of AISI 6061 T6 using Taguchi methodology and ANOVA. *J. Cleaner Prod.*, 53, 195–203, 2013.

24. Li, C. *et al.*, A method integrating Taguchi, RSM and MOPSO to CNC machining parameters optimization for energy saving. *J. Cleaner Prod.*, 135, 263–275, 2016.

25. Wang, Q., Liu, F., Li, C., An integrated method for assessing the energy efficiency of machining workshop. *J. Cleaner Prod.*, 52, 122–133, 2013.

26. Li, L. *et al.*, An integrated approach of process planning and cutting parameter optimization for energy-aware CNC machining. *J. Cleaner Prod.*, 162, 458–473, 2017.

27. Sun, Z., Li, L., Dababneh, F., Plant-level electricity demand response for combined manufacturing system and heating, venting, and air-conditioning (HVAC) system. *J. Cleaner Prod.*, 135, 1650–1657, 2016.

28. Tao, F. *et al.*, Internet of Things in product life-cycle energy management. *J. Ind. Inf. Integr.*, 1, 26–39, 2016.

29. Zhong, R.Y. *et al.*, RFID-enabled real-time manufacturing execution system for mass-customization production. *Rob. Comput.-Integr. Manuf.*, 29, 2, 283–292, 2013.

30. Zhong, R.Y., Xu, X., Wang, L., IoT-enabled smart factory visibility and traceability using laser-scanners. *Proc. Manuf.*, 10, 1–14, 2017.

31. Mourtzis, D. *et al.*, A cloud-based approach for maintenance of machine tools and equipment based on shop-floor monitoring. *Proc. CIRP*, 41, 655–660, 2016.

32. Serin, G. *et al.*, Review of tool condition monitoring in machining and opportunities for deep learning. *Int. J. Adv. Manuf. Technol.*, 109, 3, 953–974, 2020.

33. Zhou, Y., Orban, P., Nikumb, S., Sensors for intelligent machining-a research and application survey, in: *1995 IEEE International Conference on Systems, Man and Cybernetics. Intelligent Systems for the 21st Century*, IEEE, 1995.

34. Wong, Y. *et al.*, Tool condition monitoring using laser scatter pattern. *J. Mater. Process. Technol.*, 63, 1-3, 205–210, 1997.

35. Kurada, S. and Bradley, C., A machine vision system for tool wear assessment. *Tribol. Int.*, 30, 4, 295–304, 1997.

36. Haber, R.E. *et al.*, Application of knowledge-based systems for supervision and control of machining processes, in: *Handbook of Software Engineering and Knowledge Engineering: Volume II: Emerging Technologies*, pp. 673–709, World Scientific, Singapore, 2002.

37. Dong, J. *et al.*, Bayesian-inference-based neural networks for tool wear estimation. *Int. J. Adv. Manuf. Technol.*, 30, 9, 797–807, 2006.

38. Ouafi, A.E., Guillot, M., Bedrouni, A., Accuracy enhancement of multi-axis CNC machines through on-line neurocompensation. *J. Intell. Manuf.*, 11, 6, 535–545, 2000.

39. Transchel, R. *et al.*, Effective dynamometer for measuring high dynamic process force signals in micro machining operations. *Proc. CIRP*, 1, 558–562, 2012.

40. Childs, T.H. *et al.*, *Metal machining: Theory and applications*, Butterworth-Heinemann Elsevier Ltd., Oxford, United Kingdom, 2000.

41. Zeng, S. *et al.*, A novel approach to fixture design on suppressing machining vibration of flexible workpiece. *Int. J. Mach. Tools Manuf.*, 58, 29–43, 2012.

42. Salgado, D.R. *et al.*, In-process surface roughness prediction system using cutting vibrations in turning. *Int. J. Adv. Manuf. Technol.*, 43, 1, 40–51, 2009.

43. Bisu, C.F. *et al.*, Envelope dynamic analysis: A new approach for milling process monitoring. *Int. J. Adv. Manuf. Technol.*, 62, 5, 471–486, 2012.

44. Bahr, B., Motavalli, S., Arfi, T., Sensor fusion for monitoring machine tool conditions. *Int. J. Comput. Integr. Manuf.*, 10, 5, 314–323, 1997.

45. Kothuru, A., Nooka, S.P., Liu, R., Audio-based tool condition monitoring in milling of the workpiece material with the hardness variation using support vector machines and convolutional neural networks. *J. Manuf. Sci. Eng.*, 140, 11, 111006, 2018.

46. Olufayo, O. and Abou-El-Hossein, K., Tool life estimation based on acoustic emission monitoring in end-milling of H13 mould-steel. *Int. J. Adv. Manuf. Technol.*, 81, 1, 39–51, 2015.

47. Li, X., A brief review: Acoustic emission method for tool wear monitoring during turning. *Int. J. Mach. Tools Manuf.*, 42, 2, 157–165, 2002.

48. Ertekin, Y.M., Kwon, Y., Tseng, T.-L.B., Identification of common sensory features for the control of CNC milling operations under varying cutting conditions. *Int. J. Mach. Tools Manuf.*, 43, 9, 897–904, 2003.

49. Choudhury, S. and Bartarya, G., Role of temperature and surface finish in predicting tool wear using neural network and design of experiments. *Int. J. Mach. Tools Manuf.*, 43, 7, 747–753, 2003.

50. Lee, B. and Tarng, Y., Surface roughness inspection by computer vision in turning operations. *Int. J. Mach. Tools Manuf.*, 41, 9, 1251–1263, 2001.

51. Abellan-Nebot, J.V. and Romero Subirón, F., A review of machining monitoring systems based on artificial intelligence process models. *Int. J. Adv. Manuf. Technol.*, 47, 1, 237–257, 2010.

52. Xu, X. *et al.*, Intelligent monitoring and diagnostics using a novel integrated model based on deep learning and multi-sensor feature fusion. *Measurement*, 165, 108086, 2020.

53. Gosai, M. and Bhavsar, S.N., Experimental study on temperature measurement in turning operation of hardened steel (EN36). *Proc. Technol.*, 23, 311–318, 2016.

54. Byrne, G. and O'Donnell, G., An integrated force sensor solution for process monitoring of drilling operations. *CIRP Ann.*, 56, 1, 89–92, 2007.

55. Kene, A.P. and Choudhury, S.K., Analytical modeling of tool health monitoring system using multiple sensor data fusion approach in hard machining. *Measurement*, 145, 118–129, 2019.

56. Cao, Q. *et al.*, The liteos operating system: Towards unix-like abstractions for wireless sensor networks, in: *2008 International Conference on Information Processing in Sensor Networks (IPSN 2008)*, IEEE, 2008.

57. Iliyas Ahmad, M. *et al.*, A novel integration between service-oriented IoT-based monitoring with open architecture of CNC system monitoring. *Int. J. Adv. Manuf. Technol.*, 1–12, 2022. https://www.researchsquare.com/article/rs-878398/v1

4

Application of Internet of Things (IoT) in the Automotive Industry

Solomon Jenoris Muthiya[1]*, Shridhar Anaimuthu[2], Joshuva Arockia Dhanraj[3], Nandakumar Selvaraju[4], Gutha Manikanta[4] and C. Dineshkumar[5]

[1]Department of Automobile Engineering, Dayananda Sagar College of Engineering, Bengaluru, India
[2]Department of Automobile Engineering, KCG College of Technology, Chennai, India
[3]Centre for Automation and Robotics (ANRO), Department of Mechatronics Engineering, Hindustan Institute of Technology and Science, Padur, Chennai, India
[4]Department of Automobile Engineering, Hindustan Institute of Technology and Science, Padur, Chennai, India
[5]Department of Automobile Engineering, BSA Crescent Institute of Science and Technology, Chennai, Tamil Nadu, India

Abstract

Internet of Things is a method of obtaining data from several remote devices to automate and notify the user so he has a safety net protecting him from a potential failure. It is the emerging technology that can be found in every field, such as smartphone, household appliances, automated parking, commercial aviation, medical, farming, fertilizer, retail etc. We can develop a module that can be incorporated in automobile field to enhance the user experience of the vehicles, to assist the driver or to increase the comfort for the passengers, to safeguard the passengers or driver, to rescue the passengers in case of accidents, to provide secured parking, to avoid road rule violations, etc. A lot of research is ongoing in various systems of automobiles, such as engine monitoring, pollution control, accident avoidance systems, and connected cars. This chapter represents application of IoT to various systems of automobiles and the challenges in implementing IoT.

Keywords: Automobile IoT, IoT, road safety, V2X, V2V, self-driving cars

*Corresponding author: jenoris.555@gmail.com

R. Rajasekar, C. Moganapriya, P. Sathish Kumar and M. Harikrishna Kumar (eds.) Integration of Mechanical and Manufacturing Engineering with IoT: A Digital Transformation, (115–140) © 2023 Scrivener Publishing LLC

4.1   Introduction

A product that has an interface with the internet and starts sharing the local data to the host computer or a mobile phone, and provides a remote operation of the said product is called Internet of Things [1]. A dedicated system that is present online transmitting data received to a main frame. Kevin Ashton named such device as Internet of Things (IoT), he was a British pioneer and cofounder of Auto-ID Centre at the Massachusetts Institute of Technology (MIT) [2]. To get data from remote places to monitor the state of the components radiofrequency identification tags, sensors, actuators, computers, or smartphones are used [3]. IoT is the emerging technology and implemented in all fields, such as manufacturing industries for automation of the equipment. Space applications to operate devices that have been sent to outer space. IoT Health or e-health to monitor the patient's vitals to send to the family doctor, so that suffering from severe diseases, like heart attack, can be avoided. Airliners to monitor the systems and prevent failure of components [4]. The early version of this is similar to a call center where a manufacturer flight has issues, while in flight, the pilot refers to quick reference guide, and if he is unable to solve the issue, he can patch his communication system to the manufacturer call center where a trained technician can support to resolve the issues that the pilot is facing [5]. Agricultural applications, such as monitoring the soil nutrient using sensors. Moisture sensors to automatically watering the plants and to suggest a right fertilizer and pesticides for the plants [6]. IoT can improve functionality of any industries, such as manufacturing, refinery, infrastructure, oil and gas, Power sector, renewable energy, retail sector, insurance, etc. IoT can reduce the unnecessary documentation costs and wastage due to unproductive labor, improved process efficiency, and effective utilization of resources [7].

Figure 4.1 shows the applications of IoT in an automobile car. The application, connectivity type, communication type, and region segments of the global IoT automotive market. The market is split into telematics, infotainment, and navigation based on application, and embedded, tethered, and integrated based on connectivity type. The market is separated into three types of communication invehicle, vehicle to infrastructure, and vehicle to vehicle. Hardware, software, and service are the three types of offerings on the market. Asia-Pacific, Latin America, Middle East, Europe, and the North America and Africa account for the majority of the market [8]. Figure 4.2 shows the global IoT automotive market.

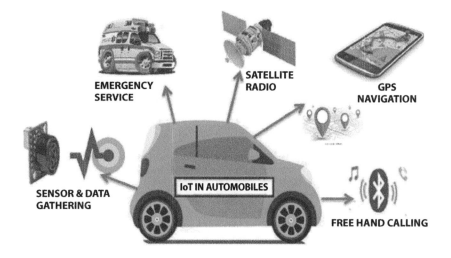

Figure 4.1 Application of IoT in automobile.

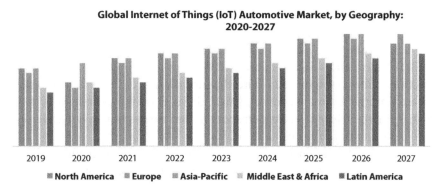

Figure 4.2 Global IoT automotive market.

The Internet of Things is sweeping the globe. It is only natural for the automobile industry to adapt and investigate the myriad of opportunity offered by IoT technology. Entertainment, security, cab services, navigation, and a slew of other features have all become standard in modern automobiles. Customer needs and satisfaction are met by self-driving autos. The vehicles, which are equipped with IoT systems, are the next generation of vehicles that are Internet of vehicles [9]. Newly launched cars are equipped with IoT to share important data, such as fuel level and expected range that the car would have, tyre pressure sensor to warn a possible puncture, remote start to cool or heat the car, etc., to monitor overall condition of the vehicle, Hundreds of sensors are equipped in a luxury cars that are

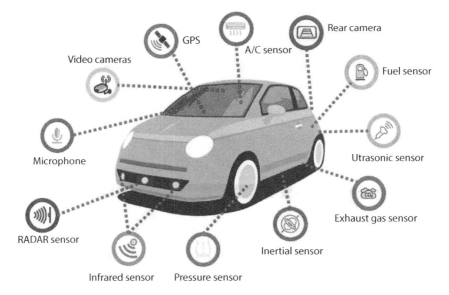

Figure 4.3 Different sensors used in car.

connected to a control unit and are connected to actuators to control the respective systems, e.g., engine management system, on board diagnostics (OBD), electronic stability control (ESC), antilock brake systems (ABS), traction control systems, etc. With the inception of IoT, one can get automated emergency management, smart driving assistance, real time fleet management, driver safety systems, connected car systems, even driverless vehicles or autonomous vehicles [10].

Figure 4.3 shows sensors used in modern car with IoT in automobiles. By this technology one can access his vehicle location and operating condition at any moment in his smartphone. The fleet owner or manager can find vehicle performance such as fuel efficiency, velocity at various times, fuel level, driver duty, halt location, goods delivery etc [11].

4.2   Need For IoT in Automobile Field

The complete IoT connected cars are expected to be deployed worldwide before 2040 and by adopting the technology people can enjoy the benefits and change the world towards a safe, secured, prompt, and environment friendly nature. One of the benefits of IoT is safety. Every year, more than 10 Lakh people are killed in road accidents [12]. Smart cars or IoT-connected cars can decrease the likelihood of accidents due to connectivity with sensors and smart apps. Collisions can be avoided by connectivity with neighboring cars with short-range wireless connectivity also known as vehicle-to-vehicle communication. Availability of another car emergency decision to the nearby cars gives a clear picture of road traffic map eases the driver to choose the right decision to reach his destination promptly. Avoiding traffic jams could save fuel. Saving fuel keeps the environments clean [13]. The connectivity among various systems of the automobile such as braking system, engine, steering system, cooling system, lubrication system, transmission system, ignition system, suspension system, etc. warns the driver well before their breakdown. So that the driver or the owner can maintain the vehicle to meet the road worthiness. Utilization of vehicle to infrastructure communication and smart technology enables the driver to get availability of nearby parking spots that cut down the fuel costs, effective utilization of parking space and time of the passengers. In case of breakdowns driver can get help through online connectivity to rectify the problem or can get information regarding nearest service stations and available facilities [14]. In case of accidents, the emergency system can send information to the nearest police stations, ambulance facilities, and also to their family members to aid and assist person involved in the accident.

4.3   Fault Diagnosis in Automobile

A great majority of current cars are more of electronic control than ever before, which is a positive thing. Because of computerization, vehicles may now be extensively inspected with self-running diagnostic checks to ensure they are in best functioning condition. These tests can look for faults or malfunctions in a car's components and integrated systems before they become a real risk [15]. A vehicle diagnostic test is a computerized analysis of the different computer systems and components in the car. Modern cars are far more digitally advanced than the average person recognises. When your car's engine is running, sophisticated software monitors different sensors and generates data reports that may later be gathered and evaluated during a diagnostic test to discover any problems [16]. Car diagnostic checks scan your car's systems and components to look for problems including the engine, transmission, oil tank, throttle, and many more. The majority of diagnostic tests are carried out by mechanics or at dealer shops since they need specialized equipment and knowledge in order to be read correctly. However, if you have the necessary expertise and equipment, you may do a car diagnostic test at home. When a "check engine" light or other warning light illuminates on the dashboard of an automobile, diagnostic tests are generally performed. In addition, they can be carried out as part of routine maintenance visits.

Current research trend shows the advancements of diagnostic techniques in the automotive industry. To name a few, a smart car health monitoring system was suggested by Bedi *et al.* [17]. Machine learning, IoT, and the AI, are discussed in relation to car monitoring, especially fuel economy, air pressure monitoring, and suspension adjustment. In addition, a variety of recently developed technologies, such as air inflation monitoring, fuel efficiency monitoring, and suspension monitoring, has been presented. The IoT-based study by Adarsh *et al.* [18] investigated the real-time monitoring of sulphur dioxide concentrations in car exhaust. This project uses the Internet of Things to identify and monitor SO_2 and CO concentrations in vehicle exhaust. Table 4.1 shows the various sensor technology, wireless protocols, and their focus in IoT.

A microcontroller processes the data and signals from three separate sensors, MQ-136, MQ-7, and DHT11, which detect and monitor temperature and humidity, SO_2, CO, concentration, respectively. Babu *et al.* [18] have developed an IoT cloud-based real-time vehicle monitoring system. GPS was used to create automobile tracking systems that make it easy for users to find their cars. The IoT based real-time automobile monitoring system can monitor vehicle parameters such as speed, engine temperature, battery level, and fuel level, as well as track the alcohol content of the driver through an Android app. When one of these measures reaches a certain limit, the system receives an alarm. Its clever RFID key safeguards your car more than traditional keys do. It is possible that the Android app will keep the car documentation, driver's licenses, insurance copies, and other papers. When an incident happens, it will send a notification to those who can be trusted about the location. Bhardwaj *et al.* [20] suggested an IoT-based vehicle health monitoring system. It is possible to learn about damage, gasoline level, engine temperature, tyre condition, and carbon dioxide emissions from a car's exhaust in this study, and you may provide comments to help avert a monetary loss. A study by Vaishnavi *et al.* [21] used IoT to develop an improved smoke monitoring system for cars. According to this research, an IoT-based car emission monitoring system will be built to track down pollution-causing vehicles on city highways and measure various types of hazardous wastes and their concentration in the air. This research proposes the use of gas sensor to create a real-time air pollution monitoring system that can be deployed anywhere and at any time. The gathered information was sent to the car's owner via text message, along with the results. This method does a fantastic job of detecting air pollution, especially in densely populated regions. Somase *et al.* [22] suggested an IoT-based system for monitoring and warning on vehicle emissions. This system keeps track of each vehicle's emission levels in real time by gathering information from

Table 4.1 Various Sensor technology, wireless protocols, and their focus in IoT [18, 19].

Inference	Gas	Wireless	Sensing technology	Strength	Weakness
System for remote gas monitoring. The information gathered is stored in a main frame and displayed on a website.	CO_2	2G	Optic	➤ Environmental characteristics are being measured ➤ GPS module ➤ On a webpage, information is shown	➤ Only one gas ➤ Periodic data transfer ➤ No IoT
For interior air quality, an energy-efficient gas sensor is used. Nodes communicate information.	CO VOC	MOS	ZigBee	➤ Low energy ➤ The presence of people awakens the sensor ➤ M2M	➤ battery based ➤ Periodic data transfer ➤ No online dashboard
Sensor gathers environmental data and sends it to an online dashboard via a gateway.	CO_2	Photoacoustic	Z-Wave	➤ Compensation for thermal ➤ No reference channel needed	➤ One target gas ➤ No application layer protocol presented ➤ Periodic data transfer

(Continued)

Table 4.1 Various Sensor technology, wireless protocols, and their focus in IoT [18, 19]. (*Continued*)

Inference	Gas	Wireless	Sensing technology	Strength	Weakness
Development and testing of a multi-gas sensor for remote air quality monitoring.	CO NO$_2$	Electrochemical	2G	➤ Environmental parameters Measured ➤ GPS module ➤ Easy and fast deployment	➤ No IoT platform integration ➤ No M2M ➤ Delay from sensed to information public
Automatic periodic calibration for a multi-gas sensor.	CO CO$_2$ SO$_2$ O$_2$ NO$_2$	Electrochemical	ZigBee	➤ environmental characteristics are measured using this device ➤ Automatic recalibration	➤ No IoT platform integration ➤ zero M2M
A multi-gas wireless sensor that uses one MCU to collect data from transducers and another to send it to the main frame.	CO CH$_4$NO$_2$ CO$_2$	Optic, Electrochemical	ZigBee	➤ Measures environmental parameters ➤ Hybrid power supply ➤ Multi-gas ➤ Small size	➤ No application layer protocol presented ➤ No M2M

several MQ sensors installed in the exhaust system. The data gathered confirms the standard limits, and this information is subsequently sent to the cloud through the IoT for further processing. If the vehicle's speed exceeds the set limit, the owner will receive a text message alarm.

Using an IoT case study for large-scale car manufacturing, Imtem *et al.* [23] worked on developing and testing an automated building energy management system. In order to enable Plug & Play installation and to configure different settings via the mobile phone, communication and control devices are required. Via addition, the aBEMS-IoT connects to a database and displays the data in a graphical interface. This study's findings showed that the aBEMS-IoT in car manufacturing worked well and saved 12% of the firm's budget. Kabir *et al.* [24] suggested work on an IoT-based parking system for the unused parking lot with real-time monitoring via an online application. Features, including automatic parking, position monitoring, parking management, invoice creation, and a payment mechanism, have been incorporated in the project. Using this approach, parking spot owners may make additional revenue by giving solutions to the general public who are always hampered by a shortage of available parking spaces.

According to Duraipaandiyaan *et al.* [25], an IoT system may be used to monitor the environment. This system uses an IoT scenario to represent the quality of the air consistently in the cloud, which can be effectively experienced from our PC or PDA. To make up for this, the system has an option to save approximated data. This aids researchers in determining whether or not the region they're investigating has an air presence, which is a useful information. Sethusubramanian *et al.* [26] development a PI-controlled converter and the monitoring of a fuel cell. This research created the concept of designing and implementing a fuel cell-based DC-DC converter and implementing IoT for remote monitoring of fuel cell characteristics. Since DC motor speed management necessitates a steady supply, the suggested work may be employed primarily in automotive applications. It can also be applied in industrial settings where DC drives are used in remote locations. MATLAB is used to simulate the circuit under consideration.

4.4   Automobile Security and Surveillance System in IoT-Based

The automobile industry is one of the fastest and largest growing industries and the real reason behind it is, more number of vehicles is on the road than 20 years before this is because easy finance led to people purchasing two cars per family, hence, the human to vehicle ratio has increased tremendously. New vehicles are entering the market and people are using them by paying a substantial amount of money. This increasing ratio of man to vehicle is increasing the crimes regarding vehicle robbery and accidents even though there are lots of safety features readily available. A simple low cost solution, based on a strong biometric mechanism that involves face authentication. The system uses a IR camera to capture the face of a person seated on the driver's seat and some sensors to provide surveillance in accidental situations. This gives us instant alerts with the latest image of the vehicle's interior on email [27]. Surveillance of on board camera. Today, most of the cars are equipped with reversing camera a device with a video buffer and a internet connectivity will work as a record of instances happen right before the accident. The buffer is programmed to record depending on the size of memory when the buffer is full it will overwrite by erasing the event that is stored 5 minutes before. This cycle repeats until the an accident triggers the vibration that will stop the recording thereby providing a surveillance footage that can be used to determine the mistake. Some luxury cars have camera on the rear view mirror to monitor blind spots, and some sport utility vehicles use multiple cameras placed on all side of the vehicle to map the terrain that the vehicle is planning to approach. All this camera are not a proprietary connect they are part of the vehicle network system by tapping into the appropriate network the feed from the camera can be accessed there by the buffer can record the feed. Figure 4.4 shows the schematic view of security and surveillance in IoT system.

Two-step authentication for vehicle access. In most of the countries the vehicle is parked on the road and not a locked garage this leads to lot of

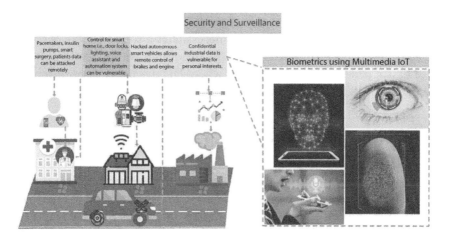

Figure 4.4 Security and survelliance in IoT system [28].

theft. This can be prevented by a two-step authentication a device in the vehicle records all the instance of door unlock and uploads the data in the main frame. Then, in order to start the vehicle, the user has to open an app in his smartphone and authenticate the entry to the vehicle. This allows the device in the vehicle to start the engine. Since both car keys and phone lost are such a rare instance. Moreover, most of the phone are protected by a biometric lock by either a finger print or face unlock, The access to the app is restricted to only to the user, hence, this two-step feature is most secure. Some of this can be automated such as commuting to the office. Since most users use to commute to certain place at the determined time with leeway, those instance alone the device lets the car start its engine without app authentication.

4.5   A Vehicle Communications

As the automobile industry evolves, a number of features are needed in the vehicle and such features are not possible by individual wires, hence in-vehicle bus protocols are developed for different applications. With a new applications in IoT, a more integral solution is needed such as getting data from the relevant data from the sensor by using the technology, which is in place than developing a special system just to make a vehicle IoT ready to enable the frequency of intra- and inter-vehicle communications [29]. A study about different protocols available for such technology and understanding bandwidth and latency in-vehicle communication bus protocols including both legacy protocols and emerging Ethernet. By tapping into the communication line, we are able to understand the messages that are being sent from various controllers and a device to decode the proprietary message and convert the data to understandable format to show up in the user's smartphone or manufacturer's computer. In the manufacturer data center a mainframe is sifting through a lot of data to find a vehicle that needs critical attention and such vehicle information is displayed on the screen of the support person, and he can take up a call to the owner of the vehicle to solve the pertaining issue. In addition to the immediate support to the customer, manufacturer can log the complaints that mainframe flags as a problem, thereby A huge reserve of such in field failures can be documented and used to rectify the issues on the upcoming products. With measurement of all driver inputs, the manufacturer can understand the common driving habits of the customer and draw out a common driving cycle can be developed to test various vehicles to see common failure points and fix the issue before the vehicle launch [30].

We have the era of big data. Traditional business to customer monitoring platforms are incapable of meeting the needs of existing users, and data costs will create invisible pressure. Data are the best assets of new transportation solution companies. With the increase in user and the accumulation of data, the interest is to discuss how to monitor the operation of data to provide effective data for operators and end users of new transportation solution vehicles. From the perspective of cluster technology, the technological innovation process of clusters and industrial clusters is complex system engineering. Under the conditions of the result of interaction, various elements of industrial clusters are formed in a specific social and economic environment to promote interaction. Second, from saving data costs to face the mechanism of industrial cluster technological innovation, theoretically analyse the value of knowledge sharing, supplementary assets and industrial cluster knowledge, and then analyse from the industrial innovation. Application mechanism and functional mechanism of the industrial cluster innovation network. Based on game theory, through the establishment of a economic model to analyse the internal mechanism of the ripple effect and integration effect under technological innovation, and verify it by measuring the ripple effect, integration effect, and technological innovation of the internal mechanism [31].

4.6 The Smart Vehicle

Smart vehicles are an important issue when the automobile industry and IoT industry fuse. After introducing the application of smart vehicles, we have more capabilities than before such as a vehicle facing breakdown failure at a critical junction or a blind turn or a narrow bridge is a hazard to other road users. So if IoT can tag the location of the failed vehicle and post a flash messages to the nearby users they can plan their approach to avoid any accident. This is very useful in particular to motorcycle riders to plan their approach. There are limitations and deficiency of this application at present and bring forward the corresponding suggestions. Meanwhile, the necessary framework which supports the applications put in practice are to be planned according to the predicted data traffic, which includes the platform which connects with other peripheral systems, the communication networks and the smart vehicle terminal [32]. This technique is also used in insurance companies to understand the customer driving patterns to make insurance payment structures that are tailored for that person and reward safe drivers with discounts.

An IoT based road surface and hazard detection and cloud update can be implemented with a simple Arduino based system with gyroscope, GPS and a GSM module can be an effective deployment while the vehicle is riding through a pothole the gyroscope measures difference in movement and records the change in cloud storage [33]. While other vehicles drive over the same surface produces data and if the data is similar it gets higher hit percentage which indicates a strong correlation with the data initially received. By setting a threshold level and the instance that gets above the threshold is reliable data and the data that gets correlation below the threshold are might be error due to local driving traffic and some changes that affect the traffic flow. After the data gets over the threshold that location along with the road condition is transmitted to the driver who will be using the road later by which the driver takes a preventive action beforehand than driving through road hazard. A simple pot hole or road irregularities can be reliably measured with gyroscope, but for other road obstructions, a camera and other sensors needed to map the exact condition, for example, a road alteration or a puddle of water cannot be detected by the gyroscope. A camera that faces the road and a machine vision system is used to determine the driver's decision as to whether he is following the road curvature or he is making an evasive maneuver to avoid certain road hazards, and a machine vision decodes the captured road surface by the camera to the normal road surface and determine the type of hazard. And also understand whether it is a constant moving hazard, such as bicyclist or a animal on the road as oppose to a puddle or a barricade blocking the road. Since this requires a significant computing power and this data is not critical, it can be processed in the cloud computer or a main frame and only transmit the data back to the other road user [34].

Most of the truck fleet operator or a delivery driver for a parcel service delivers several packages with a weight, for example, a fertilizer sack of 50 kg and several of this sacks will be a consignment to a place and other set of sacks will be a consignment like which several consignment will be delivered by a single truck. A load level sensor between the axle and the frame provides the payload on the vehicle, although it is not a calibrated load sensor, it is accurate enough to see the payload reduction when a consignment is delivered. Hence, a set of arduino, a load level sensor, and a gps can track the vehicle with the discharge load detail. An operation manager can verify the location, and the consigned load is reached by an app, and if there is an error in the delivery, he can coordinate with the driver to rectify the issue rather than the receiving party raising concerns, such as improper count or wrong delivery of goods or improper timing all can be managed through the app on a smartphone [35]. This can be also monitored by the customer placing the order to receive appropriate arrival time, so they can make arrangements to receive the goods. In some cases, the customer is in centralized location, and the delivery will be off to several sites. In such a scenario, the customer can also take delivery data via the app for payment processing. Rather than a traditional method where after delivery of the goods to site, incharge will do paper work and send it to a main office then a manager has to verify the data and finally a payment processing will be taking place. When compared to this, IoT has less labor work and quicker payment processing.

Smart Ambulance. Emergency vehicles have a blaring horn to announce other users to prioritize the vehicle than the other road occupant. This horn is psychologically affecting the patent inside the ambulance and a huge distraction. To make a smooth transitions an IoT system with a signal override and a flash message on the dash for the nearby users to instruct the decision to take in order to make way for emergency vehicle. This can be accomplished by adding a wireless IoT receiver to the on-board diagnostics port. Since the cars dash and other displays are interconnected via a network, and a simple program can allow the access to display any message the emergency vehicle operator decides to use. This form of vehicle to vehicle communication can be a huge transformation in driving. Similarly, the vehicle to infrastructure system lets the emergency vehicle to override the signals to clear the built up traffic to pass the junction with greater ease. Since the emergency vehicle is a planned route, this information shared with map providers so the map users of that local area are diverted to other route to reduce congestion in main route. This is the easiest implementation of IoT because it is a software-based feature rather than dedicated hardware based and all users posses a smartphone so the deployment is effort less.

4.7 Connected Vehicles

An IoT vehicle is one that is capable of using wireless networks to communicate to nearby cars. Connected vehicles are a crucial link about the advance of IoT in the Automotive Industry. The concept of connected cars is part of Intelligent Transportation System (ITS), which aims to provide services relating to different modes of traffic management and transport and enable users to be better informed and make safer, more coordinated, and "smarter" use of transport networks. The deployment cases vary from connected systems that connect with the driver's decision to cloud-connected vehicles that have two way communication with nonautonomous vehicles, smart mobile and four-way road signals. The most obvious and well-known use of IoT technology has been connected automobiles [36]. In the age of IoT, the term

"Information Superhighway" has taken on new significance. It is considerably more literal than the often used metaphor for pervasive networked communications. People, vehicles, and infrastructure now actually make up the information ecosystem on contemporary superhighways, owing to IoT applications that are constantly assisting in the transformation of ordinary autos into connected, intelligent, and smart transportation instruments. Connected vehicles link to a main frame to enable bi-directional data transmission between vehicles, mobile devices, and municipality vehicle control center in order to trigger critical messages and events. In the case of city traffic and intersection signals, for example, such communications can allow vehicles with connected vehicle technology to continually transmit their whereabouts and receive near real-time information.

One of the most targeted applications for the IoT in automobile is regarding safety, which may be achieved through quick intra-vehicle and vehicle-to-infrastructure unit connections (also known as V2X). V2X stands for "vehicle to everything" is the term refers to the car's communication system, in which data from a car is transmitted across high-bandwidth, low-latency, high-reliability networks, providing a way for completely autonomous driving.

In this, V2V, V2I, V2P, and V2N communications are major parts of V2X which add higher value to the future advanced mobility [36]. Automobiles will communicate with other cars, infrastructure, such as parking spaces, smartphone-wielding pedestrians, traffic signals and data centers via cellular networks in this complex ecosystem. Figure 4.5 shows the representation of vehicle connectivity in V2X technology.

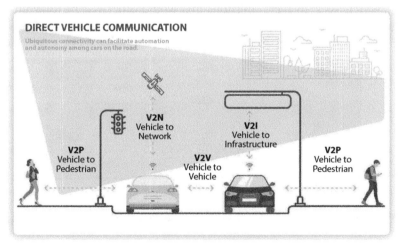

Figure 4.5 Vehicle connectivity in V2X technology [37].

The V2X technological advancements can make the driving of an individual efficient and safer compared with the lack of communication due to the reduction of the human error, which is common in every driver due to nature of the human beings.

4.7.1   Vehicle-to-Vehicle (V2V) Communications

Vehicle to vehicle (V2V) may communicate with each other in real time thanks to correspondence. This exchange is carried out via the internet using DSRC (dedicated short-range correspondence) frequencies, which are the same as those used in V2I communications. Vehicles can transmit their velocity, Gps position and heading, as well as other relevant data, providing the framework with a 360-degree view of its surroundings in V2V. Every vehicle becomes a hub that may catch, send, and retransmit signals since V2V interchanges are considered a cross-section organization. V2V is a necessary component of V2X and V2N and other V2I components. According to the National Highway Traffic Safety Administration, V2V technology might prevent 615,000 engine accidents. Because of the cross section design, cars equipped with V2V innovation have constant data on everything that happens within a 300-meter radius. In a split second, cutting-edge driving assistance systems used by a few top carmakers can use such data to alert the driver of an oncoming risk, improving street security. Figure 4.6 represents vehicle to vehicle communication in IoT wireless networks.

Furthermore, these IoT-enabled vehicles are capable of initiating preventative actions, such as timely automatic braking or suitable driver notifications, which assist to avert collisions that would otherwise be unavoidable owing to human limits [38].

Figure 4.6 Vehicle to vehicle communication using IoT wireless networks [39].

4.7.2   Vehicle-to-Infrastructure (V2I) Communications

Vehicle to infrastructure (V2I) bidirectional trade of data between the vehicle and the municipality infrastructure. This data incorporates vehicle-produced traffic information accumulated from different vehicles, information from sensors introduced in the street such as streetlamps, path markers, leaving meters, traffic signals, cameras, street signs etc, and information communicated are climate conditions, speed limits, upcoming traffic congestion, and so on. The objective of V2I is to upgrade street well-being and forestall mishaps by giving drivers continuous data with respect to various conditions out and about. Additionally, V2I and ITS advances are vital to future independent vehicles that will depend on this important data [38, 40]. Figure 4.7 shows vehicle to infrastructure (V2I) communications.

With the expanding advancement of associated gadgets all throughout the planet, the auto companies is using the accessible data to work on its items. By fitting vehicles with the V2I for sending, getting and handling relevant data, the impact on wellbeing, versatility, and comfort is creative [41]. Figure 4.8 shows the architecture of vehicle to infrastructure. The Outcomes of the V2I Communication System are as follows:

> ➤ 50% decrease in pedestrian-vehicle conflicts
> ➤ Improved mobility for pedestrian and vehicles
> ➤ Reduced emissions due to better traffic flow
> ➤ Eliminating more than 90% of road fatalities and saving $190 billion in annual healthcare costs due to accidents

Figure 4.7 Illustration of vehicle to infrastructure (V2I) communication.

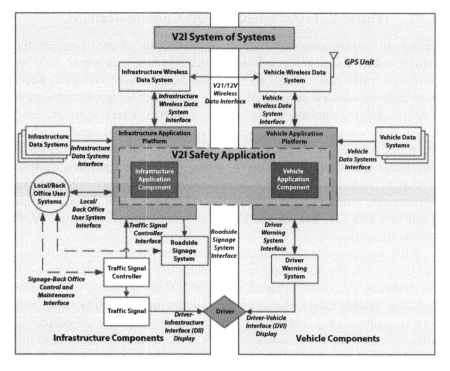

Figure 4.8 Architecture of vehicle to infrastructure.

4.7.3 Vehicle-to-Pedestrian (V2P) Communications

V2P is a part of V2X environment in which vehicle to Pedestrian (V2P) data transfer. And other V2X advancements, for example, V2V and V2P can communicate drivers intention of slowing down for zebra crossing or not can help safe crossing. The alarm cautions drivers of walkers that are drawing closer or tell the people on foot themselves of the vehicle. Without savvy street foundations (cameras, traffic signals, sensors), a stable V2P association cannot get refined. Vehicle to pedestrian means to protect the accompanying kinds of walkers while out and about:

- Children in buggies
- Individuals riding bikes
- People utilizing wheelchairs
- Passers-by
- Passengers entering and leaving public utility vehicles.

Figure 4.9 shows vehicle to pedestrian communication in intelligence transportation system. A few carmakers use in-vehicle Infrastructure, for example, LiDAR innovation to work with impact alerts, 360-degree cameras, and vulnerable side admonitions to identify walkers. Notwithstanding, the dependability of such methodologies fluctuates. This clarifies why another age of handheld gadgets and versatile applications are being created to make drivers mindful of potential crashes. V2P is carried out either directly or via network infrastructure. This will make it easier to give pedestrians alerts about approaching vehicles, as well as vehicle. The usual alerts or safety messages from the pedestrians will contain complete information about the vehicle [41, 42]. It notifies you of the direction, speed, and location of the car approaching. You can utilize this information for trajectory prediction of certain vehicles at a given time. Depending on the frequency of the vehicle's V2P, it can send about ten alerts per second (at 10 Hz).

The V2P in vehicle connectivity

1. In-Vehicle Systems to spot Pedestrians
2. Handheld Devices to alert Vehicles when Pedestrians are in the vicinity.

V2P devices that are in development:

1. Mobile Accessible Pedestrian Signal System
2. Pedestrian in Signalized Crosswalk Warning (TRANSIT)

Figure 4.9 Vehicle to pedestrian communication in intelligence transportation system (ITS).

4.7.4   Vehicle to Network (V2N) Communication

The goal of vehicle to network is to transfer data between vehicles and the management system. This procedure is made possible by a network infrastructure. Cars can receive broadcast notifications about traffic stagnation or accidents blocking the road, this information is also interpreted by autonomous driving. Vehicles can connect with the V2X management system over cellular networks thanks to vehicle-to-network (V2N) communication. V2N also interacts with other vehicles and road infrastructure via the dedicated short-range communications (DSRC) standard. Vehicles, like smartphones, tablets, and wearable devices, can now be termed "devices" with this level of connectivity [41, 42].

4.7.5   Vehicle to Cloud (V2C) Communication

To exchange data with the cloud, vehicle-to-cloud (V2C) communication uses V2N access to mobile networks to connect to the main frame. Among the applications of this technology are as follows:

➢ The software of automobiles is updated over the air (OTA)
➢ DSRC communication redundancy
➢ Vehicle diagnostics through the internet
➢ Bidirectional connection with household equipment that are also cloud-connected (IoT)
➢ Communication with digital helpers in both directions

V2C could play an important role in shared mobility. For example, during automobile sharing, riders' preferences might be stored in the cloud and used to automatically adjust content streamed to the passenger, radio stations, and other elements. The cloud providers available now are IBM IoT vehicle insights, Bosch, Cisco, Daimler, Microsoft, Amazon, Google, Harman, Ericsson, Qualcomm Technologies [42, 43].

4.7.6   Vehicle to Device (V2D) Communication

V2D communication allows cars to share data with any smart device using Bluetooth protocol such as Apple's CarPlay and Google's Android Auto, both employ this technology to link smartphones, tablets, and wearable's to a vehicle's infotainment system [44]. Figure 4.10 shows the system architecture of vehicle to device communication.

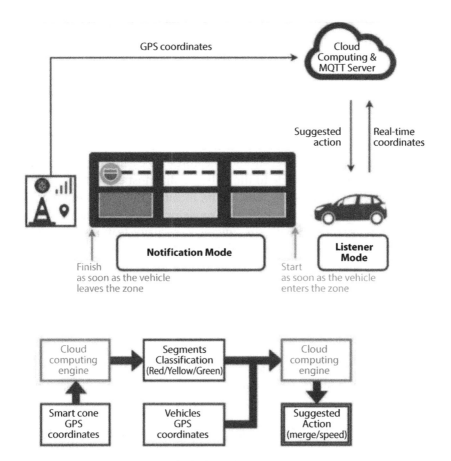

Figure 4.10 System architecture of vehicle to device communication.

4.7.7   Vehicle to Grid (V2G) Communications

(V2G) communication enables bidirectional data exchange between battery electric vehicles (BEVs), and even hydrogen vehicles (HFCEVs) and the smart grid in support of transportation electrification because V2G connectivity can help in redirecting energy for charging where required even part of hydrogen's total supply depends on electrolysis production method [45].

4.8   Conclusion

In this chapter, we have discussed the huge potential in convenient features with highest form of safety, as well as indispensable assistance whenever an accident strikes and the occupant is knocked unconscious a quick message relay to the nearest fire station can help save lives. All these features are not without a compromise, when more data relaying takes place the user privacy is at risk so until the user agrees to share the data then this is a hassle free implementation. But today's users are more concerned with sharing data to a company and its security, hence all the modalities are explained in this topic about implementation of IoT in automotive industries. We see there is a huge potential for further enhancement of the feature using IoT for the upcoming future. All these features are easily accessible just by adding a wireless IoT dongle to the onboard diagnostics port can provide many features to the existing vehicle, and it can be done on a shorter time frame. The only concern is that this dongle or the communicating main frame computer should have the advanced security features in order to prevent hackers entering with a malicious software to take control of the vehicle. This already witnessed that more and more smart features added require the robust data protection.

References

1. Nižetić, S., Šolić, P., González-de, D.L.D.I., Patrono, L., Internet of Things (IoT): Opportunities, issues and challenges towards a smart and sustainable future. *J. Cleaner Prod.*, 274, 122877, 2020.

2. Kramp, T., Van Kranenburg, R., Lange, S., Introduction to the Internet of Things, in: *Enabling Things to Talk*, pp. 1–10, Springer, Berlin, Heidelberg, 2013.

3. Jia, X., Feng, Q., Fan, T., Lei, Q., RFID technology and its applications in Internet of Things (IoT), in: *2nd International Conference on Consumer Electronics, Communications and Networks (CECNet)*, 2012, April, IEEE, pp. 1282–1285, 2012.

4. Fouad, H., Hassanein, A.S., Soliman, A.M., Al-Feel, H., Analyzing patient health information based on IoT sensor with AI for improving patient assistance in the future direction. *Measurement*, 159, 107757, 2020.

5. Pyykönen, P., Laitinen, J., Viitanen, J., Eloranta, P., Korhonen, T., IoT for intelligent traffic system, in: *2013 IEEE 9th International Conference on Intelligent Computer Communication and Processing (ICCP)*, September, IEEE, pp. 175–179, 2013.

6. Muangprathub, J., Boonnam, N., Kajornkasirat, S., Lekbangpong, N., Wanichsombat, A., Nillaor, P., IoT and agriculture data analysis for smart farm. *Comput. Electron. Agric.*, 156, 467–474, 2019.

7. Radanliev, P., De Roure, D., Cannady, S., Montalvo, R.M., Nicolescu, R., Huth, M., Economic impact of IoT cyber risk-analysing past and present to

predict the future developments in IoT risk analysis and IoT cyber insurance, 2018.

8. Maximize market research, 2022. https://www.maximizemarketresearch.com/

9. Maximize market research, 2021, https://www.maximizemarketresearch.com/market-report/global-internet-things-iot-automotive-market/10421/.

10. Huo, Y., Tu, W., Sheng, Z., Leung, V.C., A survey of in-vehicle communications: Requirements, solutions and opportunities in IoT, in: *IEEE 2nd World Forum on Internet of Things (WF-IoT)*, 2015, December, IEEE, pp. 132–137, 2015.

11. Guerrero-Ibáñez, J., Zeadally, S., Contreras-Castillo, J., Sensor technologies for intelligent transportation systems. *Sensors*, 18, 4, 1212, 2018.

12. Elvik, R., How much do road accidents cost the national economy? *Accid. Anal. Prev.*, 32, 6, 849–851, 2000.

13. Yogheshwaran, M., Praveenkumar, D., Pravin, S., Manikandan, P.M., Saravanan, D.S., IoT based intelligent traffic control system. *Int. J. Eng. Technol. Res. Manage.*, 4, 4, 59–63, 2020.

14. Taha, A.E.M., An IoT architecture for assessing road safety in smart cities. *Wireless Commun. Mobile Comput.*, 2018, 1–12, 2018. https://doi.org/10.1155/2018/8214989

15. Randall, R.B., Vibration-based condition monitoring: industrial, automotive and aerospace applications, second edition, John Wiley & Sons, 2021. https://www.wiley.com/en-us/Vibration+based+Condition+Monitoring%3A+Industrial%2C+Automotive+and+Aerospace+v2C+2nd+Edition-p-9781119477556

16. Agarwal, A.K. and Mustafi, N.N., Real-world automotive emissions: Monitoring methodologies, and control measures. *Renewable Sustainable Energy Rev.*, 137, 110624, 2021 Mar 1.

17. Bedi, P., Goyal, S.B., Kumar, J., Choudhary, S., Smart automobile health monitoring system, in: *Multimedia Technologies in the Internet of Things Environment*, vol. 2, pp. 127–146, Springer, Singapore, 2022.

18. Adarsh, K., Nair, A.S., Amruthesh, M.P., Devika, S., Krishnaveni, K.S., Jyothi, S.N., Thomas, G.M., Detection and real-time monitoring of sulfur dioxide concentration from automobile exhaust using IoT, in: *2021 6th International Conference on Communication and Electronics Systems (ICCES)*, pp. 649–656, IEEE, 2021 Jul 8.

19. Babu, M.R., Reddy, B.P., Manisha, B., Anusha, M., Vaishnavi, S., IoT cloud based real time automobile monitoring system. *Int. J. Eng. Manage. Res.*, 11, 3, 102–5, 2021.

20. Bhardwaj, A., Bala, M., Mishra, P., Gautam, J., Vehicle health monitoring system using IoT, in: *Machine Learning, Advances in Computing, Renewable Energy and Communication*, pp. 371–378, Springer, Singapore, 2022.

21. Vaishnavi, M., Nathish, P., Snehapooja, R., Madasamy, N.S., Advanced smoke monitoring system for automobiles using IoT. *Int. J. Recent Adv. Multidiscip. Top.*, 2, 3, 21–3, 2021 Mar 24.
22. Somase, A., Magdum, S.N., Vilas, H.S., Nandkishor, S.D., Vehicular emission monitoring and alerting system using IoT. *Int. J. Future Gener. Commun. Netw.*, 14, 1, 4513–20, 2021 Aug 18.
23. Imtem, N., Sirisamphanwong, C., Ketjoy, N., Development and performance testing of the automated building energy management system with IoT (ABEMS-IoT) case study: Big-scale automobile factory, in: *IT Convergence and Security*, pp. 97–107, Springer, Singapore, 2021.
24. Kabir, A.T., Saha, P.K., Hasan, M.S., Pramanik, M., Ta-Sin, A.J., Johura, F.T., Hossain, A.M., An IoT based intelligent parking system for the unutilized parking area with real-time monitoring using mobile and web application, in: *2021 International Conference on Intelligent Technologies (CONIT)*, pp. 1–7, IEEE, 2021 Jun 25
25. Duraipaandiyaan, A.P., Sathyamoorthy, S., Vishnuvardhan, M., Devi, S.S., An IoT based system for monitoring the environment. *J. Phys.: Conf. Ser.*, 1916, 1, 012162, 2021 May 1, IOP Publishing.
26. Sethusubramanian, C., Vigneshpoopathy, M., Chamundeeswari, V., Pradeep, J., Implementation of PI-controlled converter and monitoring of fuel cell on an IoT—Cloud platform, in: *Recent Trends in Renewable Energy Sources and Power Conversion*, pp. 215–228, Springer, Singapore, 2021.
27. Pawar, M.R. and Rizvi, I., IoT based embedded system for vehicle security and driver surveillance, in: *2018 Second International Conference on Inventive Communication and Computational Technologies (ICICCT)*, April, IEEE, pp. 466–470, 2018.
28. Nauman, A., Qadri, Y.A., Amjad, M., Zikria, Y.B., Afzal, M.K., Kim, S.W., Multimedia Internet of Things: A comprehensive survey. *IEEE Access*, 8, 8202–8250, 2020.
29. Soelman, M. and van der Vis, J., High-level architecture of serverless edge computing networks and its requirements. *17th SC@ RUG 2019-2020*, p. 20.
30. Uhlemann, E., Introducing connected vehicles [connected vehicles]. *IEEE Veh. Technol. Mag.*, 10, 1, 23–31, 2015.
31. Deng, Y., Agglomeration of technology innovation network of new energy automobile industry based on IoT and artificial intelligence. *J. Ambient Intell. Humaniz. Comput.*, 1–17, 2021.
32. Ma, S., The solution of an IoT application: Smart vehicle, in: *IET International Conference on Communication Technology and Application (ICCTA 2011)*, October, IET, pp. 636–641, 2011.
33. Rahman, T., IoT based smart vehicle monitoring system: Systematic literature review. *International Journal of Scientific Research in Science, Engineering and Technology (IJSRET)*, 8, 1, 12–20, 2021.

34. Lookmuang, R., Nambut, K., Usanavasin, S., Smart parking using IoT technology, in: *2018 5th International Conference on Business and Industrial Research (ICBIR)*, May, IEEE, pp. 1–6, 2018.

35. Ye, Z., Yan, G., Wei, Y., Zhou, B., Li, N., Shen, S., Wang, L., Real-time and efficient traffic information acquisition via pavement vibration IoT monitoring system. *Sensors*, 21, 8, 2679, 2021.

36. Wang, X., Mao, S., Gong, M.X., An overview of 3GPP cellular vehicle-to-everything standards. *GetMobile: Mobile Computing and Communications*, vol. 21(3), pp. 19–25, 2017.

37. Thales group.com, https://www.thalesgroup.com/en/markets/digital-identity-and security/iot/industries/automotive/use-cases/v2x.

38. Rangarajan, S., Verma, M., Kannan, A., Sharma, A., Schoen, I., V2C: A secure vehicle to cloud framework for virtualized and on-demand service provisioning, in: *Proceedings of the International Conference on Advances in Computing, Communications and Informatics*, August, pp. 148–154, 2012.

39. Extremetech, 2014, https://www.extremetech.com/extreme/176093-v2v-what-are-vehicle-to-vehicle-communications-and-how-does-it-work.

40. Syed, M.S.B., Memon, F., Memon, S., Khan, R.A., IoT based emergency vehicle communication system, in: *2020 International Conference on Information Science and Communication Technology (ICISCT)*, February, IEEE, pp. 1–5, 2020.

41. Al-Fuqaha, A., Kwigizile, V., Oh, J., *Vehicle-to-device (V2D) communications: Readiness of the technology and potential applications for people with disability (No. TRCLC 2016-06)*, Western Michigan University, 2018. https://scholarworks.wmich.edu/cgi/viewcontent.cgi?article=1030&context=transportation-reports

42. Liu, P., Internet of Thing based vehicular network system and application, in: *2018 8th International Conference on Management, Education and Information (MEICI 2018)*, December, Atlantis Press, 2018.

43. Chowdhury, D.N., Agarwal, N., Laha, A.B., Mukherjee, A., A vehicle-to-vehicle communication system using IoT approach, in: *2018 Second International Conference on Electronics, Communication and Aerospace Technology (ICECA)*, March, IEEE, pp. 915–919, 2018.

44. Vermesan, O., Bahr, R., Falcitelli, M., Brevi, D., Bosi, I., Dekusar, A., Velizhev, A., Alaya, M.B., Firmani, C., Simeon, J.F., Tcheumadjeu, L.T., IoT technologies for connected and automated driving applications, in: *Internet of Things-The Call of the Edge: Everything Intelligent Everywhere*, pp. 306–332, River Publishers, 2020. https://www.riverpublishers.com/pdf/ebook/chapter/RP_9788770221955C6.pdf

45. Haque, K.F., Abdelgawad, A., Yanambaka, V.P., Yelamarthi, K., Lora architecture for v2x communication: An experimental evaluation with vehicles on the move. *Sensors*, 20, 23, 6876, 2020.

5

IoT for Food and Beverage Manufacturing

Manju Sri Anbupalani[1]*, Gobinath Velu Kaliyannan[2] and Santhosh Sivaraj[3]

[1]Departmentof Chemical Engineering, Kongu Engineering College, Erode,
Tamil Nadu, India
[2]Department of Mechatronics Engineering, Kongu Engineering College, Erode,
Tamil Nadu, India
[3]Department of Robotics and Automation, Easwari Engineering College,
Ramapuram, Chennai, Tamil Nadu, India

Abstract

In past two decades, food and beverage industries have gone through rapid changes due to industrial revolution to meet the requirements of the 4th-generation industries. Fourth-generation industries refer to digitizing or atomizing the processes with innovative production technologies. In this context, many smart technologies have been implemented, like artificial intelligence, Internet of Things (IoT), etc. These smart technologies are based on intelligent networking of machines, allowing processes optimization and improved productivity. This chapter aims at elaborating the features of one such smart technology, i.e., IoT in food and beverage industries. The role of IoT and its implementation for water monitoring and food waste (FW) monitoring has been discussed in this chapter. Water monitoring system using IoT and the methodology for implementation of digital FW monitoring/tracking system have been discussed with its respective IoT Architectures. Thus, this chapter can lay a path for both industry people and academicians to understand the fundamentals of IoT to develop well-oriented continuous processing techniques for high productivity.

Keywords: Food and beverage industry, 4th-generation industries, Internet of Things, water monitoring system, food waste monitoring system

*Corresponding author: manjusrikongu@gmail.com

R. Rajasekar, C. Moganapriya, P. Sathish Kumar and M. Harikrishna Kumar (eds.) Integration of Mechanical and Manufacturing Engineering with IoT: A Digital Transformation, (141–158) © 2023 Scrivener Publishing LLC

5.1 Introduction

Swift and constant changes have been occurring in the food industry owing to industrial revolution, the dynamics of the industry have been changing. Digitalization has laid a path for the to use smart technologies like Internet of Things (IoT) robotics, artificial intelligence (AI), machine learning, etc., in the manufacturing processes. Also, they have laid a path for a novel phase of automation and that could enable pioneering and more competent processes, products and services [1].

The term IoT has been created by Kevin Ashton in 1999, which is a technology model intended to connect the devices and machines digitally as a vast network [2]. Thus, machines or "things" are connected via "Internet." IoT can also be referred as the Internet of Everything or the Industrial Internet. Robust communication of physical world and digital devices can be enabled by IoT, which is referred to as the fourth industrial revolution. Utilization of IoT in industry can also be referred as "Industrial Internet of Things (I-IoT)." In industries, enhanced efficiency, better decision-making and competitive advantages can be achieved by I-IoT framework, regardless of industry or company size. In I-IoT platforms, remote sensors gather data generated by machines (and increasingly, humans too), which is processed to take further decisions on operation through the network (internet protocols). Huge volume of data is collected and processed continuously within an IoT network using sensors of the various devices yield "data lake," which is possible by local physical server or cloud based storage for data processing through the suitable algorithms or machine learning techniques to generate actionable perceptions [3]. In IoT "thing" stands for a physical object for which IP address can be assigned, thereby having the potential to send data via internet protocols. For example, the insulin regulator for a diabetic patient will indicate when blood sugar level is low and activating the door lock remotely or indicating the driver when the tire pressure is low.

Based on the requirement of the case specific mechanisms of the various industries/companies, IoT can provide connectivity in three levels:

- Data is gathered from machines to monitor and displayed to operating personal directly before it goes to computer analysis.
- In optimization, computers process the collected data using analytic software to come up with actionable insights for production process. This will help the workers to take necessary manipulation in the process to improve the efficiency and productivity.

- As a result, automation come to the picture of increasing the efficiency without human interaction, by self-regulating systems. This is known as smart factories, which leads to faster responses, and significant reductions of waste and downtime.

5.2   The Influence of IoT in the Food Industry

5.2.1   Management

Transformation to a data-driven company must need improved work culture environment that revolves around the fast-moving nature of IoT to handle large amount of data. New departments and jobs must be created, such as data officers, scientists, and engineers, to handle the technology for maximum efficiency and benefit.

5.2.2   Workers

The automation through smart machines will reduce the necessity of continuous monitoring and repairs of the machines, which will reduce the number of workers. But it is beneficiary to retain the employees to deal with the advanced software reducing retrain, so as to solve problems and, productivity could be improved by employee who could make use of the additional information. Their expertise is mandatory to help for the seamless shift to a data-driven workplace based on continual improvement.

5.2.3   Data

Novel model structure is necessary for storing the data and distribute it easily. For effective data handling cloud-based analysis can be used, which is necessary for sharing the multiple locations and the status of their machines. So, any kind of factory can be implemented with best practices easily. Also, IoT can create vertical visibility in each level of the supply chain for all kind of companies, which will lead to optimization for maximum profit and minimum waste, in all the aspects of production.

5.2.4   IT

A complete integration of food manufacturing network will lead to reduction in downtime by alerting workers for maintenance needs rapidly, which provides better control in quality control during the processing, packaging, and distribution of the product. However, in order to reap the maximum benefits, all data operations in the firm must be standardized.

5.3   A Brief Review of IoT's Involvement in the Food Industry

Radio Frequency Identification (RFID) technology was used to test IoT to identify, track, and limited data storage [4]. The Internet of Things has since been applied to the development of smart technologies for many applications such as cars, homes, cities, and monitoring meters [5]. Many researchers have done a review on the Food Supply Chain tasks, which utilize IoT [6]. Also, adoption of RFID technology in the grocery supply chain have been discussed by Prater *et al.* [7]. The food sector has utilized RFID to improve the visibility of their goods in the supply chain, which complies with legal requirements [8]. Ruiz-Garcia *et al.* has done review on RFID and agricultural/food sector that utilizes Wireless Sensor Networks (WSN) technologies, as well as their applications in environmental engineering, satellite agriculture and livestock, and temperature data loggers and recording [9]. RFID was also used to track cheese-wheel movement during the production, handling, warehouse, delivery, packing, and selling phases, as well as for temperature monitoring in pineapple containers [10] and during the manufacturing, processing, storing, distribution, packaging, and marketing stages of the cheese making process, it is necessary to keep track of the mobility of the cheese-wheels [11]. The Smart Garbage Network based on IoT for effective food waste (FW) handling was also described [12].

5.4   Challenges to the Food Industry and Role of IoT

5.4.1   Handling and Sorting Complex Data

For the food manufacturing industry to be successful, it requires speed and volume with powerful analysis and optimization tools. IoT can offer such dynamics in the process via self-regulating machines. Data is read and processed by the computers and create actionable insights, which leveraged immediate action in production process.

5.4.2   A Retiring Skilled Workforce

In the food manufacturing industry, most of the present skilled workforce is nearing retirement. On such scenario, companies may use IoT to counteract this and turnover by providing innovative and sophisticated technologies in work training.

5.4.3   Alternatives for Supply Chain Management

It is difficult to conceptualize and plan properly the grid in terms of raw materials, manufacturing, and delivery of food items. As a result, technologies, such as IoT, are required by the global economy to give improved data handling and processing in order to prepare for supply roots and inventories in order to make faster and better judgments about their food items.

5.4.4   Implementation of IoT in Food and Beverage Manufacturing

Because of the extraordinary versatility of sensing devices, IoT can be used in any factory and wherever along the manufacturing chain. For the finest actionable insights that assist the industry, care must be given in handling and processing the data. Especially to be profitable and maintain the needed quality, the food sector necessitates meticulous monitoring of all systems. As a result, maintaining machine conditions and thousands of tiny systems and processes with a faultless sense of working is required to harmonize inventory with demand and assure consistent quality. The collected data from the machines can be processed by IoT and data will be centralized and placed where it is required, which leads to consistent improvement in the process efficiency of food manufacturing industries.

5.4.5   Pilot

Begin with a tiny proof-of-concept project for quick data collecting and solution, which could lead to a roadmap and plan for a complete transformation. The obtained data can be linked to a few important components of trail machinery, or the full machineries could be integrated. This technique of execution will be extremely visible to higher management and will have a good likelihood of accomplishing the goal of quick returns. As a result, thorough planning and integration takes time, and it is critical to avoid interfering too much with current infrastructure.

5.4.6   Plan

Once the pilot is successful pilot, projects must be identified with larger plan based on their priority for complete restructuring converting of a respective industry or company. As a result, investment in advanced analytic software, which is required for the third stage, must begin.

5.4.7   Proliferate

Here, changes will occur at a faster rate, necessitating the creation of new jobs, protocols, and techniques to integrate the novel process into the factory/organization. The entire process can be linked together and improved to be easily viewed in the near future.

5.5   Applications of IoT in a Food Industry

The application of IoT in the food business is endless due to the large choice of sensors and software solutions available to analyze data. The use of IoT in the food manufacturing network will change the dynamics of the process at each stage, through raw materials to completed product delivery to clients and consumers.

5.5.1   IoT for Handling of Raw Material and Inventory Control

Ensuring sufficient stock in inventory is mandatory in a food industry, which can be accomplished by IoT automatically by ordering new supplies for required product specification and quality when they are required. Optimization and complete automation will reduce waste by tracking demand. Integrated schedules with production speed will automatically order for the raw materials on the requirement, which can to minimize downtime of the process and upturn the production efficiency.

5.5.2   Factory Operations and Machine Conditions Using IoT

The built-in sensors can provide detailed information regarding machine wear and interior damage, which will make the engineers and the manufacturers to be anticipated for the remedial action plan. Predictive software alerts you to the need for repair before a machine breaks down, eliminating the need for frequent inspections and saving downtime on both counts.

5.5.3   Quality Control With the IoT

IoT may be employed to evaluate the superiority of raw materials, intermediate products, and completed goods. This will avoid from faulty goods that reaching customers. This will eradicate the need of recalling, and this supports the industry by detecting production flaws and continuously improving the operation with consistent quality.

5.5.4   IoT for Safety

A complete integrated facility is also meant for the protection and safety of the employees. Actionable insights will be gained from key safety performance indicators (KPIs), such as illness, wounds, and destruction of property. This will primarily improve staff safety and reduce losses.

5.5.5   The Internet of Things and Sustainability

Environmental sustainability could be effectively achieved by implanting IoT by aligning with guidelines automatically via digitalized tracking techniques of waste and energy usage. This will lead to usage of fewer resources to meet regulations and reduction in expenses.

5.5.6   IoT for Product Delivery and Packaging

Damaging of goods while en-route to a client can also be addressed by IoT through sensors, such sensors can sense bumping and degradation of packaging and products. Thus, the best quality is assured with lowest cost, by matching use patterns among various customers.

5.5.7   IoT for Vehicle Optimization

A novel vehicle-to-vehicle (V2V) technology is one which allows the trucks to communicate narrowly through the whole distribution cycle. New resource-saving techniques will emerge as a result of this. The ability to keep a close eye on things can save money on gas and help you avoid full stops by following traffic lights.

5.5.8   IoT-Based Water Monitoring Architecture in the Food and Beverage Industry

The IoT architecture might have multiple layers [13, 14]; but the one shown in Figure 5.1 has four: (i) the sensing layer, (ii) the network layer, (iii) the service layer, and (iv) the application layer. This structure could be beneficial for real-time water surveillance. The sensing layer that is the lowermost layer, is made up of devices or sensing devices (such as pressure transducers, temperature sensing devices, flow meters, water quality meters, and so on) whose main purpose is to acquire real-time statistics and data on water usage and quality. The network layer is responsible for transmitting and receiving data, either directly or through gateways, through a communication network (e.g., receptors and entries).

This layer tracks specified methods for interpretation of sensing devices and instruments and acts as a connection amid the database systems and software platforms of the sensing and service layers. Internet, wireless connection (Bluetooth), radiofrequency identification (RFID), Zigbee, and the Controller Area Network (CAN), and other wired and wireless networks are included. The service layer is in charge of data and information management, as well as software products and platforms. It gathers data from all IoT gateways, analyzes a massive volume of information, prior to storing it in a database system, sorting is done. Data mining and analysis is used for retrieving useful information from the stored data. Cloud-based

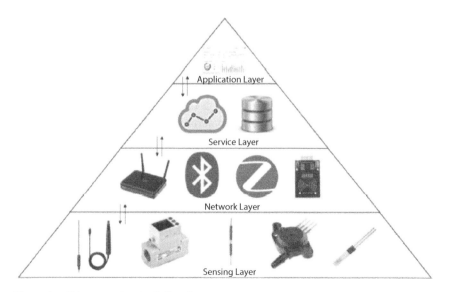

Figure 5.1 Water monitoring IoT architecture.

applications handle all of the data processing. IoT apps and services make up the application layer. It allows users to generate real-time data analysis and reports on water utilization patterns in the beverage industry, and it does so via the Internet using Hypertext Transfer Protocol (HTTP). The programs use ASP, .NET, or HTML5 to do tasks like removing wastewater out of the specific food manufacturing processes as source of raw material for further subordinate operations based on its quality relatively of distributing it to a wastewater treatment unit. It similarly aids in the surveillance of water utilization crosswise multiple manufacturing operations, sending alarms whenever any of the set values are exceeded, in addition to allowing operators to access past water utilization statistics.

Figure 5.1 shows an IoT-powered water monitoring system in a beverage facility. The selected system includes elements, such as water usage and quality monitoring, control, and regulation. The system includes a variety of components, such as pressure and water quality sensing devices, flow rate meters, interpretation of data, tools of visualization, actuators, and online and mobile control, all of which are important for communication and food production. The sensing devices are employed to determine the quality of raw water and used water, as well as their composition, water temperature, and deterioration on the pumping devices. Chlorine value, pH, electrical conductivity, dissolved oxygen, oxidation, and reduction potentinsorsal are some of the water quality metrics examined. Sensor data provides critical visions of water quality and prevailing conditions of each equipment, allowing remedial actions to be conducted if one of the parameters is outside of the defined limits. Flowrate meters aid in the measurement of real-time water use, the identification of overuse or wasting hotspots, the demonstration of proper usage, and the estimation of upcoming utilization values. The data created by the hardware (sensing devises, meters, actuators) is sent to a secure cloud server via IoT gateways (wireless routers, ZigBee, Bluetooth, etc.). The acquired data is saved on a cloud server and made available to all users in real time. The data is analyzed to discover water consumption trends, and algorithms are used to capture behavioral changes as well as differences in flow rate and water characteristics. As a sample, the water tracking network could identify leakage of water and either notify customers regarding the problem or redirect the water to storing reservoirs until the regular water supply is restored. This water tracking network was installed toward gaining a thorough understanding of the beverage factory's water utilization and, as a result, to take steps for decreasing water usage/loss, improve water quality, enhance the effectiveness of water handling mechanisms, prevent water leakages, and supervise water usage. The beverage factory benefited from the IoT-based water monitoring network in a number of ways, including increased transparency in manufacturing processes, real-time monitoring that enables for the detection and immediate resolution of water-related issues, effective human resource administration, wastewater minimization and lastly, improved decision making founded on data analytics.

5.6   A FW Tracking System Methodology Based on IoT

There are various problems in developing a FW tracking system based on IoT, and it is necessary to have the following fundamental information [15]:

1. Food waste data must be gathered;
2. The most efficient way for collecting this data must be determined.
3. The method of recording the information and communication to key decision makers
4. Analyzing and utilizing method of the information collected to eliminate the wastage of food

Grounded on the information above, finding the type of waste, and cause of generation of FW are some important criteria to quantify FW generated. As a result, implementing IoT-based FW can be done efficiently by addressing the above information, which can be used to build an IoT-based system with four layers: sensor, network, service, and application layers [16]. Because workers have complete information on FW created, this real-time computerized network can result in decrease in FW by alterations in behaviors of employees.

5.7   Designing an IoT-Based Digital FW Monitoring and Tracking System

The design of a digital food waste monitoring and IoT-based tracking system entails identifying appropriate hardware solutions that minimize human participation in data collection, as well as a software application for data storage, analysis, and transmission to key decision makers, as shown in Figure 5.1. Six steps are included in this digitized FW tracking system.

(i)   FW generated in the process line.

(ii)   Intelligent scale measurement of FW weight enabled load cells when personnel are disposing of waste.

(iii)   FW generation reasons are submitted by personnel utilizing a touchscreen-mounted software program.

(iv)   Personnel validate, confirm, and transmit FW-details to a cloud or local server.

(v)   All data generated by the FW is collected and saved.

(vi)   FW data will be investigated and submitted in user-friendly dashboards to management/staff.

This FW tracking system based on IoT employs smart scale to weigh the food waste and store it in custom software. The sort of food waste is recognized and categorized in the next phase. The specific approach uses of a human involvement to check the particulars of food waste. This data will be kept in a database utilizing cloud computing (see Figure 5.2), and then scrutinized by means of a waste monitoring system to give key decision makers with a real-time food waste tracker table.

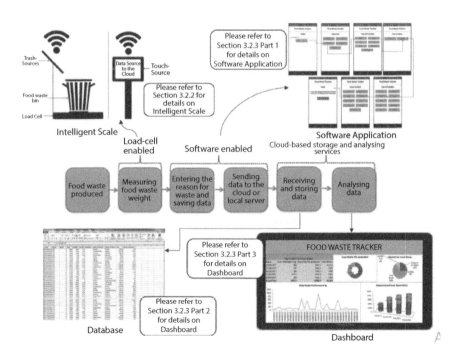

Figure 5.2 Design of IoT based FW tracking system.

5.8   The Internet of Things (IoT) Architecture for a Digitized Food Waste System

In Figure 5.3, the basic structure of the FW tracking system is shown, which consists of four strata: sensor layer, network layer, service layer, and application layer [16]. The sensor layer uses load-cell technology to determine the weight of the FW, and the software application is operated through touch screen to enter the FW cause. On the network layer, data is delivered via Bluetooth Technology to the service level using software like Arduino UNO, HM-10 BLE Breakout, and Linkit ONE. The data collected in the service layer and stored locally or in the cloud is interpreted, and more important information is extracted. The data is now available to stakeholders in the application layer.

5.9   Hardware Design: Intelligent Scale

In this step, the most relevant food waste data is collected and recorded in real time for 24 hours a day, and then analyzed using a FW tracker to provide a collection of post-interrogated data to help in the strategic reduction

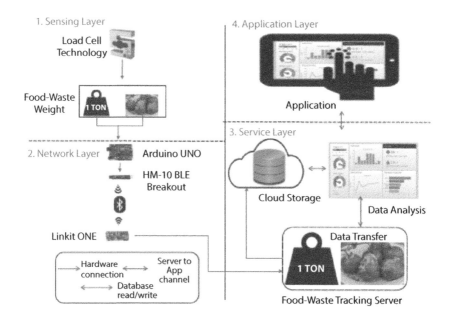

Figure 5.3 Reference architecture.

Figure 5.4 IoT-based FW monitoring intelligent scale.

and elimination of food waste. Figure 5.4 illustrates the FW surveillance system's visual representation and technology. A load cell, bluetooth adaptor, and touchscreen combo are included in the hardware. The FW bin stays on a load cell mounted platform to detect FW weight. The load cell and bluetooth technology, which might be utilized to control the touchscreen, are linked for touchscreen communication. As shown in Figure 5.4, the touchscreen comprises an application in which the employee takes a step-by-step process for entering and confirming information on the FW. This critical data about FW is provided to the FW tracking server that includes analytics and cloud storage.

5.10   Software Design

It provides stakeholders with a simple interface to view the current condition of FW and allows for analysis of FW history through time. The following are the strategic characteristics of the FW monitoring software architecture:

1. *FW application tracker*—Figure 5.5 displays the visual user interface of the FW application. To get to the result page, employees must follow basic step-by-step instructions and confirm the information by sending it to a local server or

Figure 5.5 Visual user interface of FW tracker.

the cloud. When the staff bins the tomato FW and transmits weighing information to the Touchscreen software application, the load cell underneath the bin is turned on. The example shows how readiness is linked to 8 kg of rubbish, which is worth £15.86 due to waste recycling. The date and time are added in the background once the entire information shown on the result page, which has been submitted by the staff. The trash confirmation crew's actions are depicted by the blue arrows. The only thing considered in this application was the FW created by the case company's Chicken Tikka line.

2. *Data acquisition*—During the period in which the data is collected and saved, all food waste data is acquired and kept in an excellent format in the cloud or on a local server, as shown in Figure 5.6. It includes an excellent page with all the pertinent information about food waste, including the date and time, location, type, cost, weight, and cause of FW. The Material Return Planner (MRP) tool measures weight in kilograms, with the price per kilogram me and the retail

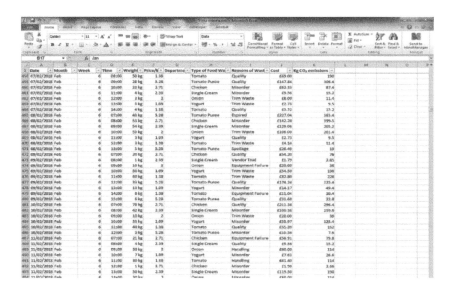

Figure 5.6 Collected data in excel format.

prices paid to suppliers is taken into account. The cost column is just multiplying FW's weight by 3.8 to get weight and price/kg and kg of CO_2 emissions (on average 1 kilograms of FW yields emissions of 3.8 kg of CO2 equivalent). Data is stored on a local server or in databases, and only the most important data is shown to investors in a user-friendly dashboard or in an exceptional style.

3. *Effective FW dashboard tracker*—The dashboard application is built by data gathered in excellent format. The application provides a real-time FW perspective of the factory floor. All procedures ensure in the development of FW monitoring, and FW data is viewed on all monitors in the plant or remotely when using the correct user ID and password. Figure 5.7 illustrates the dashboard application. Stakeholders get FW status for each food production process at predetermined intervals, such as every 30 minutes to 2 hours. Production, as seen in Figure 5.7D, is the department that produces the most waste. Figure 5.7A shows the five highest dates for highest FW throughout that month. Waste performance on daily basis (Figure 5.7E), CO_2 equivalent of the created FW (Figure 5.7B), and the source of the waste generated are some of the other key performance indicators

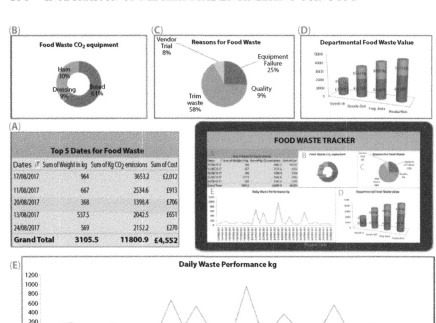

Figure 5.7 Dashboard for real-time FW monitoring.

(KPIs) related to FW (Figure 5.7C). FW data can be generated hourly, daily, weekly, and monthly for food production processes, with alerts issued if FW values vary significantly.

4. *Real-time FW Analysis*—The FW tracker compares present FW levels to preceding past best practices and gives benchmarks for additional analysis. With FW Analysis, managers can quickly identify the root cause and implement process improvements using all of the previously described FW data in real time. In addition, the problems that contribute to the production of FW may be notified by the various stakeholders to rectify the waste reduction procedure. For example, when the majority of FW is in the form of cutting waste generation by manual handling in the industrial process. The stakeholder can take better management methods to lower the FW by making this information widely available, and can integrate it into daily activities to achieve improved resource efficiency. An annual sustainability report can also be used to provide FW data.

References

1. Akyazi, T. *et al.*, A guide for the food industry to meet the future skills requirements emerging with industry 4.0. *Foods*, 9, 4, 492, 2020.

2. Rayome, A., *How the term 'Internet of Things' was invented*, TechRepublic-Louisville, Kentucky, 2018.

3. Misra, N. *et al.*, IoT, big data and artificial intelligence in agriculture and food industry. *IEEE Internet Things J.*, 9, 9, 6305–6324, 2020.

4. Ashton, K., That 'Internet of Things' thing. *RFID J.*, 22, 7, 97–114, 2009.

5. Talari, S. *et al.*, A review of smart cities based on the internet of things concept. *Energies*, 10, 4, 421, 2017.

6. Roth, A.V. *et al.*, Unraveling the food supply chain: Strategic insights from China and the 2007 recalls. *J. Supply Chain Manage.*, 44, 1, 22–39, 2008.

7. Prater, E., Frazier, G.V., Reyes, P.M., Future impacts of RFID on e-supply chains in grocery retailing. *Supply Chain Manag.: Int. J.*, 10, 2, 134–142, 2005.

8. Musa, A., Gunasekaran, A., Yusuf, Y., Supply chain product visibility: Methods, systems and impacts. *Expert Syst. Appl.*, 41, 1, 176–194, 2014.

9. Ruiz-Garcia, L. *et al.*, A review of wireless sensor technologies and applications in agriculture and food industry: State of the art and current trends. *Sensors*, 9, 6, 4728–4750, 2009.

10. Amador, C., Emond, J.-P., do Nascimento Nunes, M.C., Application of RFID technologies in the temperature mapping of the pineapple supply chain. *Sens. Instrum. Food Qual. Saf.*, 3, 1, 26–33, 2009.

11. Barge, P. *et al.*, Item-level radio-frequency identification for the traceability of food products: Application on a dairy product. *J. Food Eng.*, 125, 119–130, 2014.

12. Hong, I. *et al.*, IoT-based smart garbage system for efficient food waste management. *Sci. World J.*, 646953, 1–13, 2014.

13. Jagtap, S. *et al.*, An Internet of Things approach for water efficiency: A case study of the beverage factory. *Sustainability*, 13, 6, 3343, 2021.

14. Robles, T. *et al.*, An IoT based reference architecture for smart water management processes. *J. Wirel. Mob. Netw. Ubiquitous Comput. Dependable Appl.*, 6, 1, 4–23, 2015.

15. Jagtap, S. and Rahimifard, S., The digitisation of food manufacturing to reduce waste–Case study of a ready meal factory. *Waste Manage.*, 87, 387–397, 2019.

16. Jagtap, S. and Rahimifard, S., Utilisation of Internet of Things to improve resource efficiency of food supply chains, in: *Proceedings of the 8th International conference on information and communication technologies in agriculture, food and environment (HAICTA 2017)*, Chania, Crete Island, Greece, September 21-24, 8-19, 2017.

6

Opportunities: Machine Learning for Industrial IoT Applications

Poongodi C.[1]*, Sayeekumar M.[1], Meenakshi C.[2] and Hari Prasath K.[1]

[1]Department of Information Technology, Vivekanandha College of Engineering for Women, Trichengode, Tamil Nadu, India
[2]Department of Computer Applications, VELS University, Chennai, Tamil Nadu, India

Abstract

Machine learning (ML) plays a vibrant role in Industrial Internet of Things (I-IoT) applications and deployments. The bridge between investments and acquisitions in startups is the Machine Learning and I-IoT for the past 2 years. ML-based analytics grabs the attention of major vendors of I-IoT platform software. As the I-IoT and ML grow, there is a change in the end user responding to the market in the way industries do business. The industries and customer-oriented companies may design and define the future and will create a trend of success with these technologies. For example, majority of computer technology companies have focused toward investing in I-IoT hardware components, such as sensor nodes, actuators, to provide connectivity, and real-time data analytics. In turn, it increases the access to substantial amounts of data engendered by their customers, and also, they can use it toward improving their services and products. Previously, there was a situation in handling of data was considered as the most difficult task, but now, the scenario has changed by making availability of the data as the treasure that every company has. By the power AI, I-IoT data can be transmuted, investigated, envisioned, and implanted across the entire ecosystem, edge devices, gateways and data centers, either in the fog or in the cloud. This chapter elaborates on ML and how it could be integrated with different industrial I-IoT applications for automation and to improve businesses.

Keywords: Industrial Internet of Things, artificial intelligence, machine learning, IoT analytics

*Corresponding author: poongodi321@yahoo.com

R. Rajasekar, C. Moganapriya, P. Sathish Kumar and M. Harikrishna Kumar (eds.) Integration of Mechanical and Manufacturing Engineering with IoT: A Digital Transformation, (159–190) © 2023 Scrivener Publishing LLC

6.1   Introduction

Internet of Things (IoT) creates a new network platform in which lots of smart devices, such as sensors, mobile phones and other industry machines are inter-connected [1], which can communicate with each other and exchange information between them. Collection of data from these IoT devices and analyzing it makes the complex systems that can be constructed to enhance the quality of life, as a witness it can diagnose machine working conditions, healthcare systems,to analyze human body activities, and also structural monitoring [5]. A device or machine can be made as a smart device by sensing and communicating with other devices. The components required for making a device as a smartIoT device are shown in Figure 6.1.

IoT devices are capable of taking control without human involvement. Already IoT has rotted its power in transportation, healthcare, and automotive industries [2, 3, 6, 8]. Presently, IoT technologies are still at the research and development stage. The researchers are working on the integration of objects with sensors in the cloud-based Internet [4, 6]. In the business aspect, startups have been actively combining the IoT industry to create new services. In Lim *et al.* [7], they conducted an investigation on the IoT startup ecosystem to analyze how it is built and also to find the technologies which are transferred among startups. At the same time, with the growth of IoT, there are many challenges, such as infrastructure, the way of communications, interfaces, protocols to be used and standards, that can be incorporated.

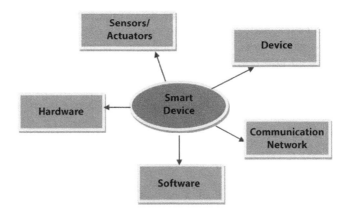

Figure 6.1 Components for making a smart device.

The IoT in assembling industry called as Industrial IoT (I-IoT) is presently basic to remain serious. This implies that turning into a computerized venture is a necessity, as Bain and Company appraises the I-IoT to have a $200B market potential by 2021. I-IoT addresses another stage in the association and control of the mechanical worth chainin turn empowering better approaches for manufacturing and constant streamlining. These are similar abilities from the I-IoT, like remote checking in Industry v4.0, which has been characterized as a name for the latest thing of computerization and information trade in assembling advancements, including digital actual frameworks, the Internet of things, distributed computing and intellectual registering, and making the brilliant production line.

In Industry v4.0 and I-IoT, the current technologies join hand with each other in the area of manufacturing and also managing the entire logistics chain, in other words it can be called as Smart Factory Automation (SFA). The Industries started relying on Industry v4.0 and I-IoT, which makes a drastic change in manufacturing by adopting agile technology and thinking innovative ways to advance production with technologies that complement and supplement human labors with robotics, which reduce unexpected accidents caused by a process failure. Regardless to ensure the consistency and process security, the transmission technology is used by engaging sensors to collect data and then process them in real time to find solutions, which support the global processes, such as asset tracking and food management, manufacturing, or medical applications. There is no other way for the industries to have global connectivity.

People are predominant in the I-IoT system, but other important measures are also shown in Figure 6.2.

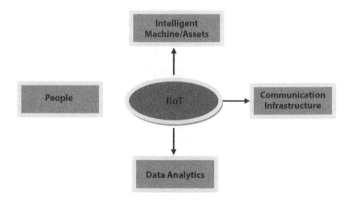

Figure 6.2 Important parts of I-IoT system.

- Intelligent Machine/Assets—In the current scenario, there is a corpus amount of data available, which is collected through sensors, motors, instruments, and other machines.
- Communication Infrastructure—In the past decades, a diversified technology called PLCs or other control systems networked in many industries, but things changed in all industries, which must be prepared itself to deal with the data deluge that comes from intelligent assets.
- Data Analytics—Data analytics is the technology which turns the data into knowledge, which created a technological revolution in manufacturing industries. This makes the people who are the creator of analytics to feel and also understand that context is important in enabling the analytics, which is useful. Analytics do not create themselves but people create them.
- People—People are the main resource in creating algorithms in spite of advent of machine learning and cloud-based predictive analytics packages that have machine learning content, like Microsoft Azure Machine Learning, IP Leanware's Brain cube, and others.

Basically, there is an understating defined by the scientist about the ML techniques is a toll for predicting I-IoT, but by its impact on wide application proved that ML is a domain and by understating its maximum benefits can be derived. All will accept that due to the influence of the ML in I-IoT, there are uncountable advantages in terms of flexibility and precision. Still, there is a challenge in handling sensor nodes, which randomly deployed in terms of link failures, energy management, memory constrains, computational factors, and also to manage the decentralization. Even though with all the above said challenges, ML stand still in solving issues, namely localization in network, clustering, data aggregation, event disclosure and processing queries, real-time routing, Medium Access Control, to maintain data integrity and in fault detection [7].

Due to the exponential growth in interface devices and sensors, the data generated is also growing rapidly. There is a need in intellectual processing of big data from real-time resources, which is directly proportional to the performance of the I-IoT application that can be achieved using machine learning techniques. Thus, the industries can provide satisfied and personalized services to their customers. To make it a successful personalized service, basically two things to must be understood properly. The first and fore most is the ability to track the behavior of the clients and the second is the ability to become accustomed to the user's changing behavior efficiently. This chapter is motivated to summarize the different ML techniques and their applications in I-IoT and also elaborates on specific tools available for implementing ML in I-IoT systems [8].

6.2   I-IoT Applications

The modern web of things (I-IoT) alludes to the expansion and utilization of the web of things (IoT) in industrial zones and applications as shown in Figure 6.3. With a solid spotlight on machine-to-machine (M2M) correspondence, huge information, and artificial intelligenge, the I-IoT enables businesses and endeavors to have better effectiveness and dependability in their activities. The I-IoT envelops industrialized mechanical applications, together with technology, clinical appliances, and programming characterized creation measures.

The I-IoT goes past the ordinary end user gadgets and internetworking of gadgets ordinarily communicates with the IoT. What makes it unique is the crossing point of data innovation and functional innovation. Functional innovation implies the connectivity of system process administration with the control systems, data acquisition, and remote human intervention,

Figure 6.3 Applications of I-IoT.

leading to the betterment of automation technology with optimum results in logistics network.

Convergence of the physics, biology, and electronics worlds paved the way for the fourth industrial revolution, enabling the industries and machines to automate their tasks and aids decision making [9].

The rally of the robots has started, with autonomous vehicles making their approach onto manufacturing floors to improve the promptness and accuracy of routine operations. These autonomous roaming robots can be coordinated in a larger level than ever before which happened with the help of the Internet of Things, by making them to be involved in automated activities in a controlled and predictable manner with minimal human monitoring. This allows them to optimize operations to a specific area, like component handling and transportation, approving them to improve efficiency, reduce risk, reduce costs, and improve data collection [11].

Automated guided vehicles and conveyors have traditionally been used to transport materials and parts throughout plants. However, the majority of these systems have relied on predetermined paths with no room for error.

Robots can now securely navigate factory floors because to the confluence of technology such as robotics, sensors, 3D cameras, 5G connectivity, software, and artificial intelligence. Some manufacturers are ahead of the curve when it comes to implementing such systems. Automotive systems manufacturer Faurecia, for example, is deploying autonomous vehicles from Mobile Industrial Robots to improve the efficiency of its operations in Italy. Workers communicate with the robots via smartphone, tablet, or computer interfaces, informing them of their responsibilities with a single button press [10].

With the appearance of the robot improvements of Industry 4.0, hobby in constructing robot grippers has elevated again. Today, the producing zone faces widening process gaps, intense deliver chain disruptions, and excessive patron expectations. New techniques to gripper layout can permit a facility to spend money on a new gadget without sacrificing facility flexibility and adaptability. New technology and layout techniques for robot grippers help make robots even greater green and bendy and accelerate a massive range of labor strategies in systems [12].

PLCs are probably the backbone of modern industry, as they allow for the consistent and autonomous execution of complicated operations. Even with PLCs, however, most automation systems require some amount of human input. The human-machine interface, or HMI, is one of the most common technologies used for this purpose, and we will look at it in this article.

HMI has a broad definition that can technically be used to any technology that allows a human to control a machine. However, the phrase is most commonly used to describe a display screen in an industrial automation system (either with touch capabilities or physical buttons). While PLCs handle the majority of the logic and low-level device connectivity, an HMI allows users to interact with the system.

The versatile feature of HMIs can enhance almost any system that uses a PLC. Most production plants, smart farms, water treatment plants, and other industrial environments use one or more HMIs. In some cases, an HMI serves as a monitoring and troubleshooting tool that passively provides warnings and other low-level details. In other cases, the entire system is designed for operators to constantly interact with the HMIs. For example, in a remote pump station, the pumps must run continuously without human intervention. An HMI at that location is only used during maintenance or when a technician is troubleshooting a known problem. However, on an assembly line, an HMI can be used to continuously control a robotic arm or other types of machines. Since an operator is always expected to be present, the HMI's ability to transmit warnings and other information in real time is much more consistent. The exact function of an HMI is highly dependent on the environment in which it is used, as well as the target user. A user interface designed specifically for engineers can look very different from a user interface designed for ordinary workers or site administrators. Fortunately, HMIs are reprogrammable and, therefore, can meet a wide variety of requirements without changing hardware.

Predictive analytics and maintenance is a methodology that guarantees cost reserve funds for routine or time sensitive preventive upkeep because of savvy metered associations. Systems are planned to help with closing the condition of in-administration hardware to estimate in what period upkeep should be executed. With the advancements in enormous information investigation and distributed computing, prescient upkeep is proceeding to develop. "Prescient support" depends intensely on information and IoT gives the best approach to breaking down information and associating all clients [10].

Intelligent logistics management is comprised of brilliant items from savvy administrations, essentially having the exact item at the precise time at a genuine spot of need and need in the administrator condition. They are imperceptible, quiet, and thus, straightforward in administration, and it sets personals free from controlling of coordinations. "Smart or intelligent" implies that arranging and booking, ICT framework, individuals and administrative policymaking should be proficiently adjusted [13].

Smart logistics is the organized relationship of these four central regions. ICT foundation gives right assets at the specific time for all the more quick and itemized data, which licenses improved arranging and booking. As we are moving into improvised knowledge, individuals and machines cooperate to serve all. Individuals are made to prepare before they involve in operating the hardware part right away. Finally, strategies, which are significant are employed, as they oversee everything in the business [14].

Wearable technology may be diligently connected with the consumer sector by the use of fitness monitors and also offers gigantic advantages in the industrial environment. In manufacturing, for example, wearable devices are progressively used to ensure professional safety by incorporating sensors to the body, which monitor environmental conditions and communicate information on vital parameters such as temperature, pulse, and respiratory rate [22].

By incorporating individual protecting equipment with sensors or radiofrequency identification technology, they become pioneering devices in the industrial Internet of Things that collect and transmit data to obtain information.

These networking platforms aim to manage workplace safety more effectively, especially in production facilities where employees perform tasks alone or handle potentially hazardous substances. It also serves to reduce administrative and compliance costs for an organization. Meanwhile, wearables are also used in production for ergonomic reasons, in order to reduce physical strain on workers through physical activities.

German automaker Audi uses ergonomic exoskeleton aids in its press shops to help workers lift and carry substantial materials. Exoskeletons also allow workers to take a sitting position if necessary.

These skills have been shown to reduce stress on the back by 20% to 30% and promote healthier posture in the long term. Such devices are becoming increasingly IoT-enabled, allowing occupational health professionals to use more accurate data for ergonomic purposes with the proper government agreements.

Improved item quality is basic for both the consumer loyalty and the business development. Components to consider for upgraded item quality is plan, creation, review, and delectability. Changes are done in item configuration to close holes distinguished and quality examination while underway with most recent advances. Faults and setbacks are identified and corrected to improvise the system value and response time. Zhang *et al.* planned a wise checking framework to screen temperature/mugginess inside cooler trucks by utilizing RFID labels, sensors, and remote correspondence innovation [18].

Brilliant stock administration connects all parts of stock administration faultlessly, from crude materials to completed merchandise. Such administration gave apparent impact utilizing structures and interconnected canny innovation frameworks continuously. There is continuous stock updates, cautions of issues, and updates of most recent cycle the stock is in and, furthermore, materials used to get yield. Radio recurrence distinguishing proof labels, standardized tags, and wise sensors are utilized to recognize, follow, and track every single required complaint and gadgets. Decrease of expiry items and improvement in work process. As an ever increasing number of actual items are outfitted with scanner tags, RFID labels or sensors, transportation and coordinations organizations can direct ongoing observing of the move of actual articles from a beginning to an objective across the whole inventory network including producing, delivery, appropriation, etc. [17].

The expression "mechanical Internet of Things" has a more muffled sounding guarantee of driving functional efficiencies through mechanization, availability and examination. Yet, the focal point of I-IoT—on industry everywhere—is more extensive as given in Figure 6.4.

Here, we take an exhaustive view, gathering together 20 top modern IoT pioneers and pioneers, drawing on the criticism from industry investigators and specialists. The spotlight here is not on sellers offering, say, a cloud-based stage for checking mechanical machines however on the organizations that themselves are utilizing modern IoT applications and innovation to drive their business forward [16].

For this component, we center around associations that utilization associated innovation pair with cloud-based investigation to drive efficiencies and dispatch new plans of action. We focus on associations that attention on coordinations, horticulture, and customary "hard-cap"

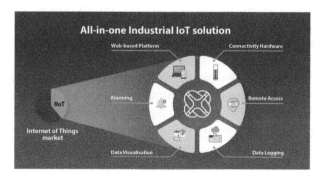

Figure 6.4 Industrial IoT solution. Courtesy: 7Devs.co.

endeavors like development, fabricating, mining, energy creation, and supply [15]. We leave out medical care, and keen city and keen structure applications, which sporadically get lumped into the mechanical iot projects area. Various organizations that have applied modern IoT are now receiving the rewards.

ABB

The most dominant company in mechanical technology firm that welcomes the idea of prescient conservationby utilizing associated sensors in robots. Its yumi model, which was intended to team up close by people, can acknowledge input through Ethernet and mechanical conventions like Profibus and devicenet. (ABB Strategy 2019 Update. Abb. Retrieved 28 February 2019.)

Airbus

In handling the Airbus, there are huge amount of parts involve in coordination where the error rate during interaction will be more. To handle this situation, Airbus incorporated Computerized Fabricating Drive called Factory of Future. The sensors and the machines, provided with the specialized wearables to be specific mechanical shrewd glasses, intended to lessen blunders and reinforce well-being in the work environment.

Amazon

The monster of the web-based retailer Amazon sets up trend by using drones for conveyance and use of multitudes Wifi associated robot Kiya in stock rooms turns media attention toward them. They grabbed this technology of using robots in tracking product in the rack most effieciently than the workers and gained 775 million dollars

Boeing

A quote by William Boeing a Flying pioneer "profits nobody to excuse any clever thought with the assertion, it is not possible." The global avionics organization formed in Boeing's name obviously still buys in to that tenet. It is presently pursuing the haggard out impartial of making its administrative contributions more momentous than its items, by dominating as the unavoidable data supplier in aeronautics. The organization is keen and taken critical steps in changing its business by forcefully innovationof IoT to drive proficiency all through is industrial plants and supply chains. The organization in addition constantly expanding the capacities of associated sensors installed into its planes.

Bosch

The indispensable motivation behind the purported Track and Trace program is that specialists would spend a sizable measure of their time chasing down devices. The organization involved the operation of sensors to its devices to track them, as a beginning with a cordless nutrunner. The intense of the Bosch turns out to be more exact and utilize the framework to direct get together tasks.

Caterpillar

Hardware producer caterpillar has been an IoT projects pioneers is deploying IoT and augmented reality applications an initial viewpoint everything from monitoring fuel levels to when air channels need supplanting. In the event that an old channel terminates, the organization can direct fundamental instructions for how to supplant it through an augmented reality application. The marine resources of the organization insight division are additionally a trend setter.

Fanuc

Provider of automation in robotics and CNC field, Fanuc employs type of robotizationthat includes sending instructions encoded into punched or attractive tape to engines that controlled the development of instruments, adequately making programmable variants of the lathe machines, presses, and milling machines.

Gehring

Gehring Technologies, an organization that makes machines for sharpening metal, was ahead of schedule to embrace I-IoT innovation. Presently, the organization empowers its clients to see live information on how Gehring's machines work before they submit a request. It does as such by utilizing computerized innovation, radiating continuous data from another machine to a client to guarantee that it meets the client's prerequisites for accuracy and productivity. Gehring utilizes a similar cloud-based continuous following to decrease personal time and streamline its own assembling efficiency through checking its associated fabricating frameworks, imagining and breaking down information from its machine instruments in the cloud.

Table 6.1 below is the partial list of companies that employ I-IoT to meet their specific requirement.

I-IoT Technology has thus become an essential component in all the sectors irrespective of the product being manufactured or service being

Table 6.1 Commercial I-IoTs.

I-IoT companies	Service provided
Augury	Detects malfunction in machinery
Embue	Construction management
Linx Global Manufacturing	Provides Automation and I-IoT tools to various industries
Noribachi	Lighting technology for industries
Valarm	Software provider for remote monitoring I-IoT applications
Smart CSM	Building - Asset management through android/iOS device
Fluidmesh Networks	Wireless connectivity while commuting
Bayshore Networks	Industrial Cyber Protection
Foghorn Systems	Provides EdgeAI solutions for Manufacturing and Energy Systems
BioInspira	Provides emissions monitoring service
NextInput	MEMs based force sensing solutions

provided—agriculture, logistics, healthcare and medical services, government, and media are the industries.

The I-IoT revolution came to reality by developments, such as network devices, communications, hardwares, data analytics, and cloud and storage infrastructure. Smart sensors, wireless networks, and gateways enable the system as powerful and affordable. The capability to communicate large amounts of data over networks for processing had enabled better analysis. This analysis of real-time data sets in offline has led to innovative business models and more effectual work processes. The harmonious interplay of all these components makes the whole greater than the sum of its parts.

6.3   Machine Learning Algorithms for Industrial IoT

In the modern advancement world of internet technologies (IT), Internet of things (IoT), and Industrial IoT (I-IoT) evidences "experiments and experience make the man perfect" is applicable to machines as well. It means that experiments and experience create machines creative and productive, which leads to the development of "machine learning." This machine learning enables machines with sensors try to replace human beings. The concept of machine learning is with an analytics engine, which processes and learns from the experience just like humans do with their brains.

Machine learning, which basically means having specialized algorithms that help system learn by itself. Machine learning is a programmed algorithm that receives the data as an input and analyze it to forecast output values within a satisfactory range. In case a new data is served to these algorithms, they will learn it by itself and also optimize their operations to increase performance and develop "intelligence" over time.

Few Applications of Machine Learning:

- Speech recognition [19].
- Classifying the Text Documents [20].
- Autonomous vehicles [21].
- Healthcare [22].
- Computation based biological application [23].
- Computer vision tasks such as image processing and face recognition [24].
- Recommendation systems, search engines, information extraction systems [25].
- Collaborative filtering [26].
- Web page ranking system [27].

To classify the machine learning algorithms based on its operation, there are four types namely

1. Supervised
2. Semisupervised
3. Unsupervised
4. Reinforcement

6.3.1 Supervised Learning

In case of supervised learning as shown in Figure 6.5, the machine will be trained. The operator will train the machine learning algorithm with a existing dataset, which includes preferred inputs and outputs, and the

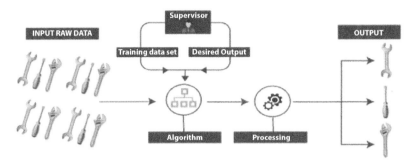

Figure 6.5 Supervised learning process in machine learning.

algorithm will analyze to attain the specified output. With the training dataset and the trainer identifies the precise answers to the problem, so that the algorithm is trained to identify patterns in data, which learns from interpretations and makes predictions. The prediction made by the algorithm will be analyzed by the trainer at the training state, and this process continues until the algorithm attains a higher level of accurateness/performance.

Further supervised learning is classified as classification, regression, and forecasting.

a. Classification: In classification, the output will be predicted based on the observed value and also the category which it belongs to.

b. Regression: In regression tasks, the relationship between the variables will be estimated. It always emphases on one dependent variable and with the series of other varyingvariables, in other words it is specifically used for prediction and foretelling in which the future is predicted based on statistical values.

Linear Regression
In linear regression inputs will be multiplied by constant to extract the output which creates the correlation between dependent variable Y and X an explanatory variable using a straight line called as regression line and its implementation [28] can be done with the general equation can for linear regression -

$$D_v = x + y \tag{6.1}$$

Where,

D$_v$ – Dependent Variable,

E$_v$ – Explanatory Variable

Support Vector Machine Regression:

As represented in Figure 6.6 in a particular training set, let {(x1, y1),....,(xi, yi)} ⊂ X × R, here X → space of input patterns. The main objective of SVM regression is to search for a fitting function f(x), the deviation should be less than Ɛ from the target (yi) attained for the relating training data set. The function should be rationallyflat or else in other way any error less than Ɛ is acceptable [30].

The linear function (f) can be represented as

$$f(x) = (d, x) + b \ with, \ db' \ X, \ b \ b' \ R$$

Here (d, x) represents the dot product of X,

Rationally flatness is described by d.

6.3.2   Semisupervised Learning

Semisupervised learning as shown in Figure 6.7, stand between both labeled, which is essential, meaningful tags, and in turn, unlabeled lacks in information.

6.3.3   Unsupervised Learning

The analysis will be done by the machine learning on data to identify the pattern. In this, no involvement will be there by the human or no answer

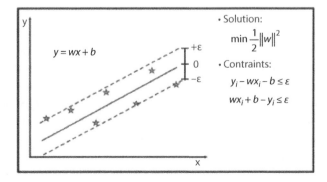

Figure 6.6 Support vector machine [11].

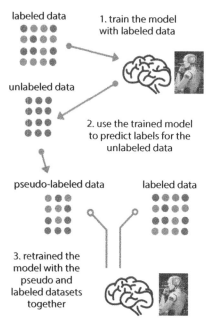

Figure 6.7 Semisupervised learning process in machine learning.

key, instead of it, the machine decides the associations and relations by investigatingon available data. Whereas in case of an unsupervised learning process as shown in Figure 6.8, it depends on corpus data sets and discourses that data accordingly. The data will be grouped into clusters or in a organized manner for prediction. Since the prediction is based on interrupting with more data, the efficiency in making conclusions on that data gradually improves and becomes more refined. It can be further classified into the following:

a. Clustering

In clustering, grouping sets of similar data based on defined criteria will be processed, which means segmenting data into different groups for execution analysis on each of it to identify the patterns. Clustering is appropriate for the examination the interrelationships between samples which is made preliminary for assessment of the sample structure. It is highly hard for human to interpret the data in a high-dimensional space, clustering is employed in tern it has real time challenges [31]. The clustering can be implemented in several ways, such as centralized, distributed, and decentralized.

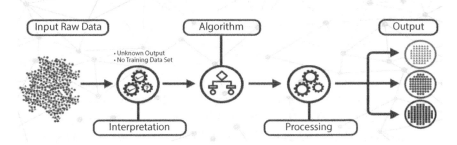

Figure 6.8 Unsupervised learning process in machine learning.

    b. Dimension reduction
       In this case, the number of variable will be reduced to iden-
       tify the exact information, which is required.

6.3.4 Reinforcement Learning

Reinforcement learning focuses on regimented learning processes as shown in Figure 6.9, here a set of actions, parameters and the end values will be provided for prediction. A set of rules will be defined in order to explore the options, possibilities, then the monitoring and evaluation process will be carried out to identify the optimal solution. Reinforcement learning process is done on the machine trial and error basis, in turn the machines learns from the past experiences to achieve the best possible result.

Reinforcement learning commonly used in industrial manufacturing [32], robotic control using AI [33], simulations for prediction [34], optimization and scheduling [35, 36], and gameplay algorithms [37, 38].

Figure 6.9 Reinforced learning process in machine learning.

6.3.5 The Most Common and Popular Machine Learning Algorithms

a. Naïve Bayes Classifier Algorithm (Supervised Learning—Classification)

The Naïve Bayes [39] classifier is constructedusing Bayes' theorem, and the values will be classified independent of any other value. This allows us to predict a class/category, based on a available set of features, using probability, and it is most efficient in text classification.

b. K Means Clustering Algorithm (Unsupervised Learning—Clustering)

The K means clustering algorithm is a unsupervised learning [40], mainly used forcategorizingunlabelled data, without any defined categories or groups.

c. Support Vector Machine Algorithm (Supervised Learning—Classification)

Support vector machine algorithms comes under supervised learning [29, 30], which analyze the data used for classification and regression analysis. It will basically filter data into categories, by providing a set of training examples

d. Linear Regression (Supervised Learning/Regression)

Linear regression [28] is the supervised leaning a basic type of regression. Simple linear regression allows us to understand the relationships between two continuous variables.

e. Logistic Regression (Supervised Learning—Classification)

In logistic regression [41], it mainly focuses on the probability of an event based on the previous data provided. The representation will be as a binary-dependent variable either 0 or 1, representing the outcomes. The main use of logistic regression is to deal with the data classifications.

f. Artificial Neural Networks (Reinforcement Learning)

An Artificial Neural Network (ANN) [42, 43] will learn by example and previous experience which is extremely applied in modeling of non-linear relationship on high dimensional data.

g. Decision Trees (Supervised Learning—Classification/Regression)

In this diagrammatic representation as a flowchart-like tree structure for branching, which examines every possibility of outcomes in Decision Tree [44, 45] in machine learning persuaded from data, rendering to a top-down approach.

h. Random Forest (Supervised Learning—Classification/ Regression)

Random forest [46] which more efficient and predominant algorithm where multiple algorithms are combined to generate attain results for classification, regression and other tasks.

Comparative study on the advantages and limitation on different machine learning algorithms is shown in Table 6.2.

On an overall, all the algorithms defined in machine language are efficient only based on the application where it is applied.

6.4 I-IoT Data Analytics

IoT data analytics platforms make it possible for businesses to evaluate and visualize sensor data from internet-based devices. They are tools used to understand the continuous flow of unstructured, structured, and time-series data generated by connected devices for organizations to understand historical data and forecast future results. It is estimated that IoT devices deployed across the world will generate 79.4 zettabytes of data by 2025, i.e., the volume of data compounds annually by 28.7%. Nowadays, IoT market is steadily growing across many fields, which in turn creates high demand to perform IoT analytics to extracting information and taking decisions from a ton of generated data.

Unlike conventional data analytics, IoT data is generally characterized by its inherent complex attributes. First, the volume of data generated by a cluster of nodes is very high. Second, the data is naturally heterogeneous and varied. Finally, by considering commercialized IoT networks alone, a rough estimation of 1.79 gigabytes of data is generated across the world every second. While many tools exist to perform IoT analytics, only a selected few can fully optimize the abovementioned volume, variety, and velocity (V^3) complexities of IoT data.

6.4.1 Tools for IoT Analytics

Below are the few top rated and widely applied tools for IoT analytics that are well appropriate to accommodate these complexities of IoT data sets. All such tools exhibited interoperability and robustness in performing analytics. They are also known for their scalability, reliability and data security to some extent in case of commercial platforms.

Table 6.2 Comparison of ML algorithms.

Algorithm	Advantages	Limitations
Artificial Neural Network (ANN)	• Efficient in detecting complex nonlinear relationships among dependent and independent variables • Needs less statistical training • Multiple training algorithms are available • Applied on both classification and regression problems	• It does not allow the user to access the précised decision making process • It is expensive to train • Pre-processing on independent variable is mandatory
Decision Tree (DT)	• It is easy to understand and interpret Data training is easier • All sort of data such as numeric, nominal, categorical are supported • Validation is can be done statistical tests	• The classes are mutually exclusive • In case the attribute value of non-leaf node is not available, the complexity raises in branching • The efficiency depends mainly on the order of the attributes or variables • The overall accuracy is less while comparing to algorithms like ANN
Logistic Regression (LR)	• Straight forward method • Updation can be done easily • No assumptions are made regarding the distribution of independent variable • Interpretation made on model parameters are efficient	• Accuracy level is not upright if the input variables relationship are complex • The Linear relationship is not taken into consideration • May exaggerate the prediction accuracy due to sampling value

(Continued)

Table 6.2 Comparison of ML algorithms. (*Continued*)

Algorithm	Advantages	Limitations
Naïve Bayes (NB)	• It is simple and very efficient with large datasets • It can be applied on both binary and multi-class classification problems • Minimal training is sufficient • It is capable of handling both discrete and continuous values using probabilistic prediction	• Classes must be reciprocally exclusive • The classification performance will be affected by the dependency between the attributes
Random Forest (RF)	• Least chance on outfitting comparing to DT • Efficient in large datasets • Its provide clear estimation on the required variable or attributes which make more impacts on classification	• It is more expensive on complex and computational aspects • Its mandatory to define the base classifiers • Occurrence of Overfitting is frequent
Support vector machine (SVM)	• Robust comparing to Linear Regression • Capable of handling multiple feature spaces • Low risk on over fitting • Performance is extraordinary classifying semi-structured or unstructured data, such as texts documents, images etc.	• Expensive on large datasets • The efficiency will degrade in presence of noise in dataset • Difficult to understand resultant model, weight and impact of variables

Google Cloud IoT Core
Google Cloud IoT Core is a completely managed platform used to manage, connect, and ingest data from millions of devices dispersed worldwide securely and efficiently. It is an IoT data analytics software that facilitates organizing global information and renders it universally applicable and accessible.

With Google IoT Core, the device registration and deployment process are surprisingly simple and straightforward. This software offers a multiple data stream option that assists in data management. All payloads possess the required information to aid in the identification and segregation of the devices.

Also, Google Cloud IoT is a scalable platform that enables users to select the best plan at the most affordable pricing. Since it is a fully managed cloud service, it greatly eliminates the troubles of any software or hardware updates. Furthermore, the availability of multiple language support like pub-sub mechanism, NodeJS, and python are beneficial for reading and controlling the data for devices.

AWS IoT Analytics
AWS IoT Analytics is another popular fully managed platform that allows the user to execute complex analytics on huge volumes of IoT data without bothering about the complexity and cost usually needed to create a customized IoT data analytics infrastructure.

By assisting to deal with a massive amount of highly invariable, inconsistent, and noisy data, AWS IoT Analytics (AIA) can help big businesses quickly obtain insight into data. This service collects and analyzes clean IoT data, which is key to the success of big companies nowadays.

Countly
It is widely known as one of the best open-source tools for IoT analytics, Countly makes a compelling market presence through its Web analytics, mobile analytics and marketing platform capabilities. It is developed on Node.js, Countly's open-source SDKs are compatible with a range of modern-day devices—Web-based, mobile tech, smart TVs, smart watches, and other IoT smart devices.

Billions of data points from across several devices are processed on the cloud to generate customizable reports. Countly provides real-time data dashboards with a maximum time latency of up to just ten seconds, which is on par with various costly enterprise IoT analytics options available in the market.

In Countly, Web analytics is provided from a granular level. User profiles, attribution analytics, campaign tracking, session frequency tracking, geolocation (city/country) tracking, crash reports are a few of the many detailed insights provided by it. The tool also gives users the option to create funnel visualization and heat maps.

Things Board

This open-source IoT platform for collecting, processing, analyzing, and visualizing telemetry sensor data is scalable, fault-tolerant, and geared for high-performance computing. The toolkit supports both on-premise and cloud deployments.

The toolkit's core service, the Things Board node, written in Java, is responsible for transferring data using REST API calls. Fully customizable, Things Board clusters offer possibilities to create a range of technical microservices—HTTP/MQTT/CoAP transport microservices, WebUI microservices and JavaScript executor microservices. Rule-based data processing algorithms can be applied to normalize, validate, or transform input data sets. Users can also customize the rule engine toolset from the Things Board dashboard to drag/drop Rule Nodes or define Root Rule Chain.

Things Board is best known for its real-time IoT dashboard. The toolkit offers more than thirty customizable widgets to create rich visualizations, perform deep analytics, and provide compelling IoT use cases.

Thing Speak

The data aggregation and analytics IoT toolkit, Thing Speak, offers non-commercial open-source solutions that can visualize IoT device data using MATLAB widgets. Thing Speak's reputation can be attributed to its seamless integration with the MathWorks product suite. This robust IoT analytics tool supports RasberryPi, Arduino, and Nodemcu devices.

IoT sensor data transferred to the Thing Speak cloud using restful APIs and HTTP protocols can be analyzed and visualized for more in-depth insights using MATLAB software. There are also options to retrieve data in JSON, XML, and CSV formats for manual data analysis and reporting. There are options for users to share data with their teams using private and public channels. Thing Speak also has a paid commercial toolkit, but its open source and free-to-use solutions that work alongside MATLAB computational algorithms are more than well suited for performing the fundamental IoT data analysis and visualizations.

Apache Stream Pipes [47]

This industrial analytics toolkit is known to help both nontechnical and technical users to collect, analyse and study IoT data sets. Stream Pipes uses machine learning algorithms to perform advanced analytics, pattern detection, predictive analysis, anomaly detection, and temporal analysis. It is well-reputed with nontechnical users thanks to the intuitive, easy to use Web interface, and graphical editor.

Stream Pipes Connect, the in-built channelization framework, can collect data inputs both from IoT device archived and real-time data sets. Stream Pipes also comes with built-in semantics to provide intelligent insights and recommendations for data stream elements and/or transformational modules. The toolkit is compatible with HTTP/REST, MQTT, Kafka, OPCUA, and ROS protocols. Enterprise key process indicators (KPIs) and production reports can be visualized in real-time using Web based cockpits.

There are additional options for software developers to use wrappers like Apache Flink and Apache Spark to customize SDKs and Maven archetypes to create new data processing elements. One of the unique features of Stream Pipes is its ability to aggregate geographically distributed data pipelines in real-time, thereby creating possibilities to perform edge computing on IoT data. Various data harmonization algorithms, like filters, aggregation and unit converters, help developers clean and enrich device sensor data, periodically.

WSO2 IoT Server

A server for the IoT platform released under Apache 2.0 license, this toolkit is trusted to offer versatile solutions with edge computing. WSO2 IoT Server creators pride themselves on its seamless integration, easy-to-deploy drag/drop widgets and platform scalability. The platform can manage up to a million IoT devices and provide deep data analytics of all the data aggregated from them.

WSO2 uses WSO2 Data Analytics Server (WSO2 DAS) to perform real-time analysis, batch analysis, interactive analysis and predictive analytics. WSO2 Complex Event Processor (WSO2 CEP) is used to handle millions of data aggregations per second. This makes the analytics platform well suited to processing enormous volumes of IoT data.

WSO2 also provides analytics extension event adapters for HBase, Rabbitmq, and Twitter, in addition to the native built-in event adapters available on its analytics platform.

Previously, the processes of collecting, storing, and analyzing an enormous volume of data sets was considered a complex and expensive task.

But, today, with IoT standardization, cloud computing, machine learning and edge computing, IoT analytics is taking huge progressive leaps in the industry. Our commercial world may not be fully adapted yet to leverage the power of IoT analytics, but we are definitely getting there.

Industries, such as retail, pharma, healthcare, manufacturing, and even smart city projects, are increasingly gaining momentum in terms of artificial intelligence, machine learning and data analytics. Fortune 500 companies are already deploying IoT architecture in order to have a better understanding of their business processes and customer preferences. IoT data analytics is transforming enterprises and businesses already.

Datadog
This is the analytics, security, and monitoring platform for business users, security engineers, IT operation teams, and developers in the cloud age. As a SaaS platform, it automates and integrates infrastructure monitoring, log management, and application performance monitoring to deliver coherent, real-time observability to the entire technology stack of their customers.

Businesses of all sizes are utilizing Datadog and across a vast range of industries to facilitate cloud migration and digital transformation, enable collaborative efforts among development, security, business, and operations teams, fast track time to market for applications, better understand client behaviors, secure infrastructure, and applications, alleviate time to problem resolution, and monitor crucial business metrics.

AT&T IoT Platform
AT&T IoT data analytics platform offers fully managed cloud services, as well as multinetwork connectivity support for developers of IoT solutions. Also, it transforms and optimizes a massive volume of data while also taking care of security at every level of your team when using their services.

Bella Dati
Bella Dati is an agile IoT data analytics and reporting software that allows business users to use real-time business data to make informed decisions. It is a pure web application, which means that it requires no installation. It also features a lot of data source connectors.

Oracle Internet of Things Cloud
Oracle IoT offers the broadest range of cloud solutions to power innovation and simplifies IT. Users commend Oracle Internet of Things for its ease of use and straightforward dev-friendly implementation.

6.4.2   Choosing the Right IoT Data Analytics Platforms

When considering these IoT data analytics platforms, there are certain things user should be looking for examining depending on the solution are as follows [48]:

Privacy/Security
It is a must to consider how the users have resolved privacy and security issues in the past and evaluate their security provisions. It would be best to examine how their software handles security challenges and how it shields the user away from the complexity.

Data plan
Also, consider whether the vendor offers a reasonable data plan. Also, whether the data services have to be able to suspend or pause as it demands, as well as the ability to determine how much data is being used.

Service type
How does the IoT vendor present and sell their services? Some platforms are entirely connectivity platforms, while others are end-to-end services that provide the connectivity, software, and hardware required. It would be best if the vendor considered what the business needs are. How likely are your needs going to vary over time?

Market Experience
It should also be taken into consideration how long the IoT provider would be operational. Although the IoT market is relatively new, it is, however, rapidly growing. It is recommended to find an IoT provider that has been active in the sector for 4 years or longer.

Mode of connectivity
A brief analysis has to be done to address what kind of connectivity the application required and whether a cellular mode is enough or a Wi-Fi solution is required by IoT system for its optimal performance. It must examine these needs and consider how the vendor will address them.

Connectivity
It is pertinent that the user also considers how the vendor's network coverage will fit the business's present and future initiatives nicely.

OTA firmware updates
Examine how the vendor enables you to send updates and find solutions to bugs on your systems remotely. Is it going to be an easy or complex procedure?

Device management
Consider how the vendor allows you to manage, segment, and monitor IoT devices that are already out in the field.

Hardware
Enquire if the vendor offers any starter packages, developer kits, or off-the-shelf applications for the particular use case the business is targeting. While the user will probably have the need to do a couple of customizations, the user will save significant resources (time and effort) by not beginning from scratch.

API Access/Managed Integrations
It is to be understood how the IoT data analytics platforms integrate all the sophisticated components required for our IoT business or solution. These components can include RTOS, application layer, security, cloud connections, firmware updates, device diagnostics, carrier/SIM cards, cellular modems—into a unique package that will ease the workload for engineering the IoT platform.

6.5 Conclusion

This chapter has dug out the potential for ML to treadle I-IoT to support the realization of tracking and analyzing the services in Industries through the use of ML tools. Industries always rely on information technology because of the rapid growth of IoT significantly in recent years. This leads to better energy efficiency, good quality products, reduced costs, improved decision-making potential, and less equipment downtime and other customization. The adaptation of these technological updates should be handled carefully, as there are still appropriate concerns related to consistency, cost-effectiveness and safety. Many transformations are essential to take place to mark ML and I-IoT worthwhile in the industries. Most significantly, software and hardware need to be wangledtogether to address these innovative technologies and their role in the Industrial applications. The capability to understand the processing of the machines for given environments and to be able to adjustrequired services to make this to fit better with their needs efficiently is fundamental for any successful industry application. Overall, the study indicates that a blend of ML and I-IoT offers a great potential to developers for providing various smart industrial services.

References

1. Tang, C.-P., Huang, T.C.-K., Wang, S.-T., The impact of Internet of Things implementation on firm performance. *Telemat. Inform.*, 35, 7, 2038–2053, 2018.
2. He, W., Yan, G., Da Xu, L., Developing vehicular data cloud services in the IoT environment. *IEEE Trans. Industr. Inform.*, 10, 2, 1587–1595, 2014.
3. Singh, V.K., Kushwaha, D.S., Singh, S., Sharma, S., The next evolution of the Internet-Internet of Things. *International Journal of Engineering Research in Computer Science and Engineering (IJERCSE)*, 2, 1, 31–35, 2015.
4. Hepp, M., Siorpaes, K., Bachlechner, D., Harvesting wiki consensus: Using wikipedia entries as vocabulary for knowledge management. *IEEE Internet Comput.*, 11, 5, 54–65, 2007.
5. Guinard, D., Trifa, V., Karnouskos, S., Spiess, P., Savio, D., Interacting with the SOA-based Internet of Things: Discovery, query, selection, and on-demand provisioning of web services. *IEEE Trans. Serv. Comput.*, 3, 3, 223–235, 2010.
6. Joshi, G.P. and Kim, S.W., Survey, nomenclature and comparison of reader anti-collision protocols in RFID. *IETE Tech. Rev.*, 25, 5, 234–243, 2008.
7. Lim, S., Kwon, O., Lee, D.H., Technology convergence in the Internet of Things (IoT) startup ecosystem: A network analysis. *Telemat. Inform.*, 35, 7, 1887–1899, 2018.
8. Aceto, G., Persico, V., Pescapé, A., Industry 4.0 and health: Internet of Things, big data, and cloud computing for healthcare 4.0. *J. Ind. Inf. Integr.*, 18, 100129, 2020.
9. Balaji, S., Nathani, K., Santhakumar, R., IoT technology, applications and challenges: A contemporary survey. *Wirel. Pers. Commun.*, 108, 1, 363–388, 2019.
10. Bayoumi, A. and McCaslin, R., Internet of Things – A predictive maintenance tool for general machinery, petrochemicals and water treatment, in: *Sustainable Vital Technologies in Engineering & Informatics*, 2016.
11. Breivold, H.P. and Sandström, K., Internet of Things for industrial automation – Challenges and technical solutions, in: *2015 IEEE International Conference on Data Science and Data Intensive Systems*, IEEE, Sydney, 2015, 10.1109/DSDIS.2015.11.

12. Jaidka, H., Sharma, N., Singh, R., Evolution of IoT to IIoT: Applications & challenges, in: *Proceedings of the International Conference on Innovative Computing & Communications (ICICC)*, 2020

13. Kawa, A., SMART logistics chain, in: *Intelligent Information and Database Systems. ACIIDS 2012*. Pan, JS., Chen, SM., Nguyen, N.T. (eds), Lecture Notes in Computer Science, vol. 7196, Springer, Berlin, Heidelberg, 2012. https://doi.org/10.1007/978-3-642-28487-8_45

14. Kim, D.S. and Tran-Dang, H., An overview on industrial Internet of Things, in: *Industrial Sensors and Controls in Communication Networks*, pp. 207–216, Springer, 2019. https://doi.org/10.1007/978-3-030-04927-0

15. Leminen, S., Rajahonka, M., Wendelin, R., Westerlund, M., Industrial internet of things business models in the machine-to-machine context. *Ind. Mark. Manage.*, 84, 298–311, 2020.

16. Niranjan, M., Madhukar, N., Ashwini, A., Muddsar, J., Saish, M., IoT based industrial automation. *IOSR J. Comput. Eng. (IOSR-JCE)*, 2, 8, 36–40, 2017.

17. Radanliev, P., De Roure, D., Nurse, J.R., Mantilla Montalvo, R., Burnap, P., Supply chain design for the industrial Internet of Things and the industry 4.0. Preprints 2019, 2019030123.

18. Xu, L.D., He, W., Li, S., Internet of Things in industries: A survey. *IEEE Trans. Industr. Inform.*, 10, 4, 2233–2243, 2014.

19. Deng, L. and Li, X., Machine learning paradigms for speech recognition: An overview. *IEEE Trans. Audio Speech Lang. Process.*, 21, 5, 1060–1089, 2013.

20. Nguyen, T. and Shirai, K., Text classification of technical papers based on text segmentation. *Natural Language Processing and Information Systems*, pp. 278–284, 2013.

21. Chen, Z. and Huang, X., End-to-end learning for lane keeping of self-driving cars. *2017 IEEE Intelligent Vehicles Symposium (IV)*, 2017.

22. Kononenko, I., Machine learning for medical diagnosis: History, state of the art and perspective. *Artif. Intell. Med.*, 23, 1, 89–109, 2011.

23. Jordan, M., Statistical machine learning and computational biology. *IEEE International Conference on Bioinformatics and Biomedicine (BIBM 2007)*, 2007.

24. Siswanto, A., Nugroho, A., Galinium, M., Implementation of face recognition algorithm for biometrics based time attendance system. *2014 International Conference on ICT For Smart Society (ICISS)*, 2014.

25. Thangavel, S., Bkaratki, P., Sankar, A., Student placement analyzer: A recommendation system using machine learning. *4th International Conference on Advanced Computing and Communication Systems (ICACCS-2017)*, 2017.

26. Wei, Z., Qu, L., Jia, D., Zhou, W., Kang, M., Research on the collaborative filtering recommendation algorithm in ubiquitous computing. *2010 8th World Congress on Intelligent Control and Automation*, 2010.

27. Yong, S., Hagenbuchner, M., Tsoi, A., Ranking web pages using machine learning approaches. *2008 IEEE/WIC/ACM International Conference on Web Intelligence and Intelligent Agent Technology*, 2008.

28. Huang, M., Theory and implementation of linear regression. *2020 International Conference on Computer Vision, Image and Deep Learning (CVIDL)*, pp. 210–217, 2020.

29. Deshmukh, P.R. and Borhade, B., Support vector machine classifier for research discipline area selection, in: *2017 International Conference on Intelligent Computing and Control Systems (ICICCS)*, pp. 462–466, 2017.

30. Byun, H. and Lee, S., Applications of support vector machines for pattern recognition: A survey. *Pattern Recognition with Support Vector Machines*, pp. 214–215, 2002.

31. Tamilselvi, P. and Kumar, K.A., Unsupervised machine learning for clustering the infected leaves based on the leaf-colours. *2017 Third International Conference on Science Technology Engineering & Management (ICONSTEM)*, pp. 106–110, 2017.

32. GaoYang, Z.R., Hao, W., Zhixin, C., Study on an average reward reinforcement learning algorithm. *Chin. J. Comput.*, 30, 8, 1372–1378, 2007.

33. Jurgenschmidhuber, J.Z. and Wiering, M., Simple principles of metalearning. *Rock Soil Mech.*, 31, 155–156, 2010.

34. Fu Qiming, L.Q., Wang, H., Xiao, F., Yu, J., Li, J., A novel off policy Q(A) algorithm based on linear function approximation. *Chin. J. Comput.*, 37, 3, 677–686, 2014.

35. Z.M. and Wei, Y., A reinforcement learning-based approach to dynamic jobshop schedulin. *Acta Autom. Sin.*, 31, 5, 765–771, 2005.

36. Ipek, E. and Mutlu, Self-optimizing memory controllers: A reinforcement learning approach. Presented at the *2008 International Symposium on Computer Architecture*, 2008.

37. Tesauro, G., TD-Gammon, a self-teaching backgammon program, achieves master-level play, in: *Neural Computation*, vol. 6, no. 2, pp. 215–219, March 1994.

38. C.S. and Kocsis, L., Bandit based monte-carlo planning. *ECML*, 282–293, 2006.

39. Huang, Y. and Li, L., Naive bayes classification algorithm based on small sample set. *2011 IEEE International Conference on Cloud Computing and Intelligence Systems*, pp. 34–39, 2011.

40. Sinaga, K.P. and Yang, M., Unsupervised K-means clustering algorithm. *IEEE Access*, 8, 80716–80727, 2020.

41. Yang, Z. and Li, D., Application of logistic regression with filter in data classification. *2019 Chinese Control Conference (CCC)*, pp. 3755–3759, 2019.

42. Mishra, M. and Srivastava, M., A view of artificial neural network. *2014 International Conference on Advances in Engineering & Technology Research (ICAETR - 2014)*, pp. 1–3, 2014.

43. Qiang, W. and Zhongli, Z., Reinforcement learning model, algorithms and its application. *2011 International Conference on Mechatronic Science, Electric Engineering and Computer (MEC)*, pp. 1143–1146, 2011.

44. Esposito, F., Malerba, D., Semeraro, G., Kay, J., A comparative analysis of methods for pruning decision trees. *IEEE Trans. Pattern Anal. Mach. Intell.*, 19, 5, 476–491, May 1997.
45. Jijo, B.T. and Abdulazee, A.M., Classification based on decision tree algorithm for machine learning. *J. Appl. Sci. Technol. Trends*, 2, 01, 20–28, March 2021.
46. Patel, S.V. and Jokhakar, V.N., A random forest based machine learning approach for mild steel defect diagnosis. *2016 IEEE International Conference on Computational Intelligence and Computing Research (ICCIC)*, pp. 1–8, 2016.
47. Open Source For You, https://www.opensourceforu.com.
48. IoT World Today, https://www.iotworldtoday.com.

7

Role of IoT in Industry Predictive Maintenance

Gobinath Velu Kaliyannan[1]*, Manju Sri Anbupalani[2],
Suganeswaran Kandasamy[1], Santhosh Sivaraj[3] and Raja Gunasekaran[4]

[1]Department of Mechatronics Engineering, Kongu Engineering College,
Perundurai, Tamil Nadu, India
[2]Department of Chemical Engineering, Kongu Engineering College,
Perundurai, Tamil Nadu, India
[3]Department of Robotics and Automation, Easwari Engineering College,
Ramapuram, Chennai, Tamil Nadu, India
[4]Department of Mechanical Engineering at Velalar College of Engineering &
Technology, Thindal, Tamil Nadu, India

Abstract

The notion of the Internet of Things (IoT) refers to the eloquent networking of smart gadgets over the network of the internet. Industries/organizations are now experimenting with various ways of predictive maintenance, such as a way of cutting down the expenses and minimize the intervals between maintenance procedures. Because they can combine data from many machines and production systems, IoT platforms are useful support for predictive maintenance. The communication framework is the biggest barrier in integrating production systems with IoT specialized platforms, as most routing protocols of industrial communication are inconsistent for using newer communication routing protocols employed on IoT platforms. A broad overview of current PdM concerns is given in this work, with the goal of better understanding of advantages and disadvantages, difficulties, and scenarios of the concept of dynamic maintenance. This is accomplished by doing in-depth research and analysis of scientific and technical publications. On this foundation, this chapter addresses some key research concerns that must be solved in order for IoT-enabled PdM to be developed and used successfully in industry. A case study is also discussed partly for establishing the viability of

*Corresponding author: gobinath.v.k@gmail.com

R. Rajasekar, C. Moganapriya, P. Sathish Kumar and M. Harikrishna Kumar (eds.) Integration of
Mechanical and Manufacturing Engineering with IoT: A Digital Transformation, (191–214) © 2023
Scrivener Publishing LLC

suggested technology for enabling persistent assessment of high-end machines via battery-enabled IoT equipments. The deployed test bed, which is made up of 33 devices that gather data through IoT. This interprets data from temperature analysis and vibration assessment activities, especially for 2 months. The evaluation of data transfer delay and prediction of working life with the aid of power consumption.

Keywords: Internet of Things, predictive maintenance, challenges, battery-enabled IoT devices

7.1 Introduction

As the industrial processes shift toward predictive manufacturing, maintenance plays important role as a value generation function within manufacturing must be modified in order to achieve more sustainable operations. New maintenance strategies are offered in this area [1]. The so-called Internet of Things (IoT) vision has been promoted in recent years by ongoing improvements in electrical industries, also the creation of novel highly superior and economically feasible communication via wireless methods [2]. Knowing that failure prediction is a critical technology for ensuring long-term operations, many fault prediction systems based on IoT are now being developed [3].

IoT is defined as the smart networking of advanced equipments that allow items to perceive and communicate with one another, so altering the conditions and authority of making decision about the physical environment. Initially, the idea of IoT has been deployed for the products such as household appliances and controlling devices of smart home, which connects IoT via cell phones or network devices [4]. Other sorts of items were equipped with the requisite technology as the concept grew, and a lot of industries have been attempting for integrating the idea of IoT with technologies for contemporary production. The main stage in developing a generic, cloud-based, predictive maintenance system, which simplifies factory upkeep, is gathering useful information and values of various process equipment utilizing the network of IoT. Connecting ordinary industrial equipment and network platforms is the biggest disadvantage of employing IoT technology in production operations. The current research provides a new way for translating the data from process equipment to IoT compatible one via technologies, such as Arduino or Raspberry-Pi microcontrollers, to address this issue [5–8].

Interoperability refers to capacity for the communication and utilization of data between several IT devices. It relies on standardized procedures and level of work assessment. Whether a business creates its IoT system in-house or hires a provider, it is critical that creating a viable and long-lasting solution. Internet technologies such as HTTP GET and HTTP POST, as well as the Java Script data-interchange format (JSON), are used to communicate with other devices and IoT systems. Considering that all of those communication systems are not supported on industrial equipment, using Arduino as a translator among industrial equipment and IoT platforms is a quick option [5] (Figure 7.1).

Industrial Internet of Things (IIoT) solutions can be utilized to successfully develop smart factories that achieve improved levels of efficiency. IoT smart objects can be widely utilized for collection of required data for attaining advanced technology in manufacturing sector [10], increasing worker safety irrespective of their position [11], and minimizing the occurrence of faults [12]. Cutting-edge and persistent predictive maintenance solutions are constructed by employing sensing devices for monitoring the process condition of each equipment, lowering maintenance costs, and averting harmful circumstances. Furthermore, by considering IoT devices that can connect and collaborate with one another, any delays caused by manual intervention are eliminated, and a quick response for

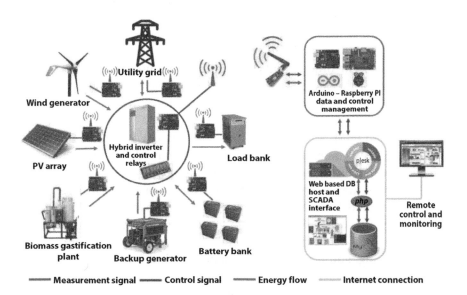

Figure 7.1 Simplified scheme of an IoT platform with Arduino board [9].

crucial situations is attained. Numerous communication protocols have to be employed that allow equipment collaborate with one another in order to achieve remote monitoring technologies that are both powerful and intuitive in the Industrial IoT environment.

7.2   Predictive Maintenance

Predictive maintenance (PdM) is the most recent technique that many sectors have embraced. Power plants, public services, transportation, and emergency services are all examples of industries that require perfect reliability. Forecasted data is typically required for long-term planning and numerous operational actions (maintenance, production, inventory, etc.). Furthermore, maintenance cannot always be conducted everywhere due to technological and logistical constraints [13]. Maintenance is an important part of the manufacturing process. Machine failures in the middle of a manufacturing run might cause delays in delivery or require employees to work more hours to make up the loss. To optimize maintenance efforts, PdM anticipates system failures [14, 15].

PdM refers to instruments employed that assess whether it is necessary to do particular maintenance according to Carvalho [16, 17]. The technology has attributed to constant monitoring of the equipment or process, allowing to do the maintenance , as when it is required done. A secondary, but no less significant, role of PdM is the ability to detect defects early, using methods based on historical data (machine learning) and visual features of faults (color and wear). PdM intends to cut maintenance costs, achieve zero-waste production, and reduce the number of significant failures as part of the Industry 4.0 idea [18]. Despite the benefits of PdM, Herrmann [19] warns against the dangers of remote access to maintenance procedures, citing DDoS attacks as an example. We differentiate three approaches to PdM, according to Zonta [8], namely: Based on a physical model, the key element of which is mathematical modelling, which necessitates state timeliness and statistical methods of evaluation.

The second way is the knowledge-based approach, which minimizes the physical model's complexity, and the third approach is the data-driven approach, which we see most frequently in current PdM development. This method is based on artificial intelligence, i.e. machine learning and statistical modelling, and it is a suitable method for Industry 4.0 [20]. Farooq *et al.* differentiate between experience-driven and data-driven maintenance [21]. Preventive maintenance based on experience is focused on acquiring information about production equipment and then using that information to plan future maintenance. Data-driven preventative maintenance, on the other hand, is based on analyzing a significant amount of data (Figure 7.2).

Predictive maintenance refers to series operations, which identify variations in process parameters and equipment's condition (signs of failure) so that the required maintenance work can be performed to extend the equipment's service life while reducing the risk of failure.

According to the ways of identifying failure signals, it is divided as:

1. Statistical
2. Conditional

Condition-based predictive maintenance relies on monitoring the state of machineries a regular basis to detect failures will lay a path to come up with maintenance decisions, whereas statistical-based predictive maintenance (SBM) relies on statistical data from the meticulous recording of stoppages of in-plant items and components to develop models for predicting failures. Our clever predictive maintenance solution (Figure 7.2) provides businesses with exciting new options. Through data mining systems, data collected by CPS are communicated via IoT, and the process parameters are simultaneously evaluated for any patterns that suggest a potential malfunction. This decision makes use of IoTs to allow for early detection of a stoppage and the most efficient implementation of corrective procedures. It also implies that unplanned downtime may be avoided, and

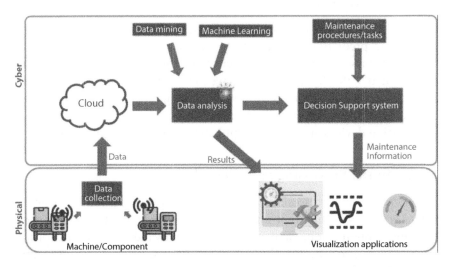

Figure 7.2 The architecture for integrated predictive maintenance (IPdM) technologies [22].

people and resources can be better utilized. For intelligent predictive maintenance, this new approach is known as Industry 4.0 (IPdM).

7.3   IPdM Systems Framework and Few Key Methodologies

IPdM methodologies rely on a variety of important approaches, including CPS, IoT, IoS, computational intelligence (CI), data mining (knowledge discovery), and swarm intelligence (SI), is investigated and established in order to meet business needs.

IPdM consists of six primary modules:

1. Detection and collection of data
2. Initial processing of collected data
3. Modelling as per requirement
4. Influential parameters
5. Identification of best working path
6. Modifying output with respect sensed input

The corresponding modules were created in NTNU's KDL laboratory that can be explained as follows.

7.3.1   Detection and Collection of Data

Its initial step of adopting the methodology of IPdM for machine monitoring and prediction is to complete the phase. This module's job is to find the right sensor and the best sensor method. The sensor signals are transformed to the domains, which have the maximum data about the state of the machinery throughout the data collecting procedure. Microsensors, ultrasonic sensing devices, vibration sensing devices, and acoustic emission sensing devices are built for capturing several types of information.

7.3.2   Initial Processing of Collected Data

In general, two processes are there for dealing through sensor signals. Signal is processed first that improves signal's quality and characteristics. Filtering, amplification, data compression, data validation, and denoising are all signal processing techniques that increases signal-to-noise ratio. Therefore, the next one is feature extraction that extracts characteristics of an impending failure or problem from processed data. In general, three domains can be used to extract features: time domain, frequency domain, and time–frequency domain. In IFDPS, altogether, the approaches can be selected and employed, thereby determined by a real-world equipment or system analysis.

7.3.3   Modeling as Per Requirement

The maintenance decision-making module provides adequate useful data for maintenance, employees to make informed decisions about maintenance tasks. There are four types of decision-support models to choose from: There are four types of models: (1) physical, (2) statistical, (3) data-driven, and (4) hybrid. Since the majority of IPdM approach is based on signals and data that indicate equipment status, the data-driven model will take the lead. IPdM is a data-driven and hybrid model company. In the maintenance decision-making module, IPdM focuses on data-driven and hybrid methodologies.

If historical data is readily available, data-driven analysis is an excellent way to pinpoint a problem and assess its severity. When just a portion of the historical data is available, hybrid methodologies that mix data-driven and model-based methods are utilized for efficiently assessing state of the equipment. When just a portion of historical data is available, the semi-supervised learning approach may be used to evaluate condition and find faults, and it is quite successful. All of these strategies can be chosen based on the examination of a genuine production system. Diagnostics and prognostics are the two primary types of techniques used in the maintenance decision-making module.

When defects arise, fault diagnostics focuses on detecting, isolating, and identifying them. Prognostics, on the other hand, tries to forecast defects or failures before they happen. Equipment diagnosis is increasingly using CI and DM techniques, which have demonstrated to be more effective than older methods. However, because to a lack of effective approaches for obtaining training data and particular expertise necessary to train the models, it is difficult to deploy CI techniques in reality. Until now, the majority of applications in the literature have relied solely on experimental data for model training.

Artificial neural networks, fuzzy logic systems, fuzzy-neural networks, neural-fuzzy systems, evolutionary algorithms, and swarm intelligence are among the intelligent approaches employed. The number of articles on prognostics is substantially lower than that of diagnostics. The most common prognosis is the amount of time until a failure occurs. The remaining usable life is the term used to describe how much time is left (RUL). IPdM assesses the RUL using a data-driven model then attempts for discovering relationships between RUL and machine or component's state.

7.3.4 Influential Parameters

A KPI diagram, often known as a spider chart or a health radar chart, can be employed to show component deterioration. Each radio line indicates state of the machine, which ranges as zero (excellent) to one (defective). Also, colors represent corresponding state of the component, which include safe, warning, alarm, fault, and damage. The graphic can aid workers and managers in visually evaluating the equipment's performance.

7.3.5 Identification of Best Working Path

Maintenance planning also schedule optimization can be the type of NP issue, and SI algorithms might be a useful strategy for solving it. IPdM seeks to discover the best dynamic predictive maintenance scheduling using Genetic Algorithm (GA), Particle Swarm Optimization (PSO), Ant Colony Optimization (ACO), and Bee Colony Algorithm (BCA). In IPdM, all of these strategies may be used to tackle maintenance scheduling optimization challenges.

7.3.6 Modifying Output With Respect Sensed Input

Based on the results of the maintenance decision-support module, this module can perform error correction, compensation, and feedback control.

7.4 Economics of PdM

It makes sense to develop IoT-based PdM in place of Industry 4.0 that is proven as lucrative as conventional maintenance methods. To assess the economic benefits of PdM, cost of maintenance models is constructed. However, despite the importance of this issue for the choice to engage in IoT for PdM, only a few initiatives have been undertaken in this direction [23–26]. However, instead of establishing universal analytical methodologies, these efforts rely on simulation.

Additional cost-benefit metric is presented as the technical value (TV) [27] that is taken into consideration of the performance during critical failure mode detection, diagnostics, and prognostics, as well as expenses accompanying with false alarms [28–30], though TV has cost parameters, which are challenging for quantification (for example, the savings obtained through segregating a problem ahead of time) and it uses constant performance measures, i.e., metrics that are not dependent on time (e.g., the probability of a failure mode). Finally, TV fails to account for incorrect detection, diagnosis, and prognosis. Partially Observable Markov Decision Processes (POMDPs) [31, 32] are employed for evaluating the value of information of the data acquired through sensing devices deployed upon civil structures while taking into account the unpredictability of condition monitoring. In a nutshell, VoI has high price and decision-maker pay for knowledge that could be useful collecting only when its value exceeds its cost [33].

Despite the fact that this framework appears to be highly promising, there are two major difficulties that prohibit it from being used for Industry 4.0. Specifically, the VoI can be defined as projected savings, which is obtained via lowering the ambiguity of deterioration state assessments using data gathered by sensors or even inspections, i.e., VoI is used to choose exploratory and inspection activities [31, 32]. Then there is the fact that it is a relative number, which makes it difficult to compare PdM to other techniques that are not focused on sensor monitoring. As next is difficulty that limits the implementation of the VoI technique is time consumption of the algorithms used and only appropriate for small-scale scenarios with a small number of state-action combinations. Within a real choices framework, a model to assess the value of PdM at the system level is developed and suggested. PdM has been viewed as an investment tool for the Decision Maker (DM) for alternatives and future maintenance activities by developing a cost–benefit–risk model [34].

Few of the challenges persist to employ in industrialised practise, such as calculating the cost difference between doing CM Vs RUL-driven maintenance. Furthermore, while time-dependent RUL predictions have been considered, they are not tied toward predictive algorithms' efficiency (accuracy, precision, etc.), therefore, the progression of economic variables associated to RUL predictions into the options model is described using a Brownian motion process [34]. Furthermore, the model only examines CM as a feasible option to PdM than any other preventive maintenance options; yet, here are circumstances where CM and SM outperform CBM and PdM in terms of economic performance [35].

Refined analytical methods have been established to ensure the cost-benefit analysis of canary-based PHM to maximize component resilience is described as a set of reliability and restoration, with the latter is a function of PHM character traits, thereby a life-cycle maintenance cost analysis framework is evolved that takes into account time-dependent false and missed alarms for fault diagnosis, and where time-variant metrics from the research works have been interconnected; and these theoretical techniques do not adequately represent the dynamism of the CBM situation, where a choice must be made each time the PHM algorithms are performed [36–41]. The improvement of economic models is a prerequisite for the sector to release IoT investments for PdM.

7.5   PdM for Production and Product

When we analyse about the suitability of PdM for any equipment, we must first differentiate between a scenario which considers PdM for a product and the case where it is applied to equipment employed for manufacturing. The first scenario, cost-benefit explanation is straightforward: a number of companies from various industrial sectors (e.g., manufacturing [42], aviation [43–45], mining [65], energy [46], etc.) are interested in PdM since it improves commercial competitiveness. Furthermore, new sources of income can be generated through new chances for added value in service, such as taking on a portion of the clients' business risks and other (financial) burdens: the prospect of new business may be sufficient motivation for investing in PdM on its own.

The usefulness of PdM and, by extension, the IoT infrastructure, is more difficult to quantify in manufacturing operations. Production process in the automotive industry can be used for demonstration, also drawing some broad conclusions. Figure 7.3 depicts the flow of a manufacturing process with multiple steps separated by buffers. In general, for smaller buffers, application of the just-in-time paradigm is stricter. In further process steps, the value of half-processed steps is higher, then, the buffer values are smaller. Therefore, initial manufacturing steps (i.e., production of shell via welding, milling, and so on) can be much more hopeful from a PdM standpoint, as they are carried out for capital-intensive areas of the plant, using robots, transportation systems, welding equipment, etc.

Instead, in last phase (assembly), which is highly time-sensitive, screwdrivers, travelling cranes, and other machines with relatively high redundancy and manpower levels are used. The value of PdM in this system is strongly function of the buffers, which has a Bf level for enduring an upstream production step downtime of D hours has been calculated through Eq. 7.1.

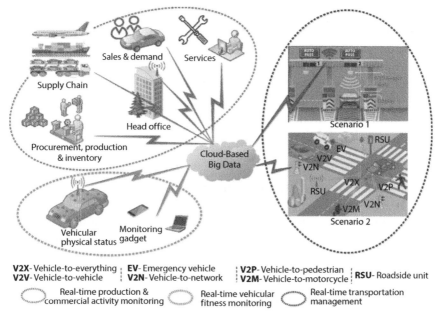

Figure 7.3 IoT-integrated automotive industry [47].

$$\text{Bf} = \frac{1}{takT} \times D \qquad (7.1)$$

where takT—time duration between initial production point of one unit and next unit (in h)

D—downtime (in h)

To arrive at suitable numbers, we can conservatively assume D = 10 h of downtime, then takT = 6 min = 0.1 h of downtime; this gives us Bf = 100. The corresponding mobilised capital cost was determined using Eq. 7.2.

$$\text{Mc} = C_p \times \text{Bf} \qquad (7.2)$$

Where,

Mc—mobilized capital (in units of cost)

Cp—product cost at final stage of fabrication process

If C_p = 3000 euro, for example, Mc = 3000000 euro. PdM must pay 300000 euro each year to minimize business interruptions, anticipating a capital cost as10% per year. This is a fraction of the cost of a large-scale IoT investment for PdM. Then, only if we can predict the various indirect costs of failure, like costs to refill the buffer, warehouse charges, costs associated with conservative scheduling of scheduled maintenance intervals, which has outcome of over maintenance expenses, etc., Therefore, the abovementioned expenses can validate the significance of PdM spending. For IoT investment justification, this underscores the requirement for robust cost models encoding PdM.

7.6   Implementation of *IPdM*

Projects in European were completed, Norwegian and Chinese enterprises and the educational institutes using Industry 4.0: IPdM solutions [48]. Some of the relevant problems encountered were mentioned as follows.

7.6.1   Manufacturing with Zero Defects

The EU-funded project Intelligent Fault Correction and Self-Optimizing Manufacturing System (IFaCOM) aims to develop a predictive maintenance approach in smart factories.

7.6.2   Sense of the Windsene INDSENSE

Norwegian national research project, developed a system that lowers unpredicted operational shutdown of wind turbines, which boost sup time of power plant [49].

7.7   Case Studies

An expanded test bed is established for power plant to evaluate the performance of the described system in real-world settings. Three monitoring networks, in particular, have been deployed in three different sections, which are interlinked with RCSR via system gateways. All the sections were selected by consulting employees, with the goal of identifying operational systems where continuous temperature and vibration monitoring might improve maintenance procedures. Areas selected are as follows, 1st area is heavy ash evacuation, 2nd area is seawater pumps, and 3rd area is evaporators. All the instruments, machines, and shop floor activities are recorded in the following sections.

7.7.1   Area 1—Heavy Ash Evacuation

Seven sensing instruments are installed in chosen spot to measure vibrations and temperature in numerous spots of an ashes water pump. Because all of the gadgets are powered by powered, the entire area was fitted shortly and the gateway installation taking the most time. Since this is not powered by battery, it is also interlinked with power plant's main network via a wired connection. Because all magnetic solutions were chosen by plant staff involved in monitoring chores, no additional time is needed to sensing devices and probes. Installation of four nodes to monitor vibration three nodes to temperature is next done nearby vital elements on machine that is deemed to be of exceptional importance and is very substantial in size.

Figure 7.4 depicts a number of nodes that have been placed. The entire monitoring region has been controlled via single gateway, whose location was determined using experiments using packet loss rate (PLR) measurements from the furthermost sensing devices. PLR is smaller than 10% for most of the chosen jobs. Because most of the gadgets are powered by battery, the entire installation was done in minimal time, except the gateway installation that has taken the most time. Because the respective components are not powered by battery, but they are interlinked with power plant's main network via a wired connection. Because all magnetic solutions were chosen by plant staff involved in monitoring chores, no additional time requirement was there in sensing probes placement in the machine itself.

7.7.2   Area 2—Seawater Pumps

Two pieces of machinery, feeding pump and a lubricating oil exchanger, have been chosen to be monitored in the second area. Four sensors have been mounted on the feed pump and further three sensing devices was installed on the lubricating oil exchanger. In the second section, totally seven sensor devices was deployed. Because two machines are separated by around 10 meters, with various impediments in the way, and the gateway's sole suitable position did not provide satisfactory PLR scores, three nodes of the respective routers were built. The router was essentially can be the sensing device that has no sensing capability and is interlinked with power supply. The principal function was aimed at sending data packets as per the RPL protocol's established routing patterns, allowing industrial IoT systems to fully utilize multihop communications. The final PLR throughout the entire area is less than 5%, thanks to the utilization of the three routers.

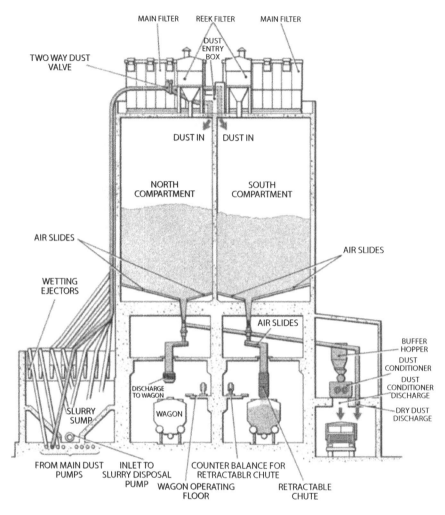

Figure 7.4 Layout of Ash evacuator (station design and layout—British Electricity International, https://doi.org/10.1016/B978-0-08-040511-7.50009-7).

The entire area was erected in one and a half days, with some delays owing to the need for additional electrical wire to power the routers. Magnetic probes were also used in this situation due to their ease of placement.

7.7.3   Evaporators

Temperature and vibration sensing devices were fitted in several pumps in this region, resulting with four pieces of equipment ejector, conden-sate, distillation, and recirculation pumps) being monitored by 19 wireless sensors. In this section, during one and a half days of labor, the entire monitoring devices and a gateway were put. The site of the gateway has been determined by many field tests, also dependent of PLR values, necessitating extra duration for the area's entire installation. As a result, gateway installation was the way that it can ensure a maximum PLR of less than 10%. A few of sensing devices of area 2, as well as pumps with temperature and vibration sensing devices: the ejector also the condensate. The magnetically mounted and clamp-mounted sensing devices were also employed in this example.

7.7.4   System Deployment Considerations in General

The employed systems are architecturally identical with higher end design, which is evident along with extra one monitoring section. This scalable system, then new sections was simply included additionally, since network administration and data storage capabilities will be the enhanced in gateway connected sections. Commissioning duration of the three locations was around within a week time, and deployment of electrically powered gateway and routers was one of reasons for time delay. A sensing device take only a few minutes to install, and it is fortunate for features of the IPv6 protocol, also instantaneous recognizing capability of the system (i.e., play capabilities and plug), as well as all entire details of sensing device (exposed resources, IPv6 address, etc.) is displayed by remote control and service room graphical interface. In installation phase, the only setup action necessary is to associate the sensor with the monitored machinery.

7.8   Automotive Industry—Integrated IoT

IoT has a high magnitude of relevance as an intelligent technology, as evidenced by numerous automotive industries and expected future usage [30]. IoT has wide application in continual and periodic assessment of vehicle condition through immense data acquisition, data processing and interpretation that are encased within the IoT paradigm and used for a variety of reasons.

7.8.1   Navigation Aspect

Importance of an IoT-based car movement in simulation and practical applications imposes serious impact in providing safer travel to the vehicle passengers and drivers [50]. This includes useful characteristics such as displaying of destination routes through GPS irrespective of area for safer driving assistance. When a vehicle encounters emergency challenges such as an accident, an IoT-enabled model employing the SKM53 GPS module and the Haversine formula can alert the nearest rescue team for immediate assistance and speedy aid to the victims [51]. In addition to self-driven vehicles, IoT has some applications in high end vehicles tracking, vehicle monitoring and passenger safety assessment [52, 53].

7.8.2   Continual Working of Toll Booth

Managing traffic in optimistic approach enhances the country's economic stability [54]. Due to a lack of effective traffic control, a typical vehicular motion might occur, resulting in serious disasters, cluttering the road, and affecting national economic growth. A smart traffic management system has the potential to alleviate these problems [55]. Further increase of vehicle movement, AI-enabled traffic maintenance through sensed data through deployment of RSUs [56], and CCTVs [55]. In an emergency, the central system can assist in locating the accident site and alerting the nearest rescue squad [54]. It also informs the driver about road conditions (asphalt, wet, and snow), traffic, and other unintentional information on the road, as well as emergency parking locations [57]. Intelligent traffic management may be traced all the way back to 1992, when the first RFID technology application was used to collect road toll charges [58]. IoT technology now aids in the improvement of this system by offering security, a comfortable working environment, and improved traffic management through traffic congestion reduction and hassle-free payment [59]. Through a web/database query technique, the system can also determine the vehicle's physical features before imposing a toll charge [60].

7.8.3   Theft Security System

Theft alerts in vehicles is a standard method of preventing theft. These capabilities are now enhanced with the use of IoT technology, making the car more cost-effective, secure, and dependable [61].

7.8.4   Black Box–Enabled IoT

Black Box is a device used to collect and consolidate the data captured at the time of disaster of aero-vehicles. During operational mode, Black Box is deployed as an electronic footage gathering device to record various facts for future examination. Unfortunately, in many earlier situations, the black box could have been destroyed permanently owing to unforeseeable fatal accidents, and as a result, investigators were unable to uncover the true cause of road accidents. The system can capture images, record video, collect location coordinates, and perform data processing utilising analysis tools with the support of IoT technology. It subsequently sends the necessary data to a cloud server, where it is shared with the chosen responsible person through email and SMS [62]. In recent years, EDR has been necessary for installation within the vehicle to record incidental information for postaccident forensic reports [63].

7.8.5 Regularizing Motion of Emergency Vehicle

Traffic jam is a major issue all over the world, and it can result in the death of individuals irrespective of availability of ambulances at the right time to provide timely service [63]. By giving alarm to other vehicles, shall provide alertness to other riders travelling in the same road [64].

7.8.6 Pollution Monitoring System

As per Health Organisation, over 2.4 million people die each year as a direct result of air pollution [65]. Emission of numerous un-burnt gaseous particles from vehicles and industries, air pollution has become a major issue. Conventional automobiles were major root cause for the environmental air pollution, accounting for anything from 12% to 70% of total emissions. IoT may also be used to monitor air pollution levels at various locations across the country [63].

7.8.7 Timely Assessment of Driver's Condition

Vehicle drivers should be aware of public safety such as pedestrians, opposite vehicle drivers and passengers [66]. Due to poor life style, the physical and mental conditions of drivers get deteriorated. Hence, timely assessment of driver's health is necessary. This can be done using the sensors integrated with advanced IoT leads to the prevention of accident occurrence [67].

7.8.8 Vehicle Performance Monitoring

Based on the elements affecting the performance of vehicle, the optimal working of vehicle can be achieved through incorporation of IoT. Static analysis and monitoring was too easy process, as compared to the dynamic monitoring process. The engine will not perform at elevated temperature, further leads to higher fuel requirement and declined compression ratio of turbocharger [63]. Hence, it is very important to make a note on each and every components of vehicle. This might prevent the occurrence of serious accidents. The futuristic goal of IoT incorporated vehicular monitoring system were increased comfortless, reduced fuel combustion, improved mileage, reduced vehicular vibration, etc., IoT, as a developing technology, has a lot of room to grow in the automobile industry, thanks to its advanced gadgets and communication technologies.

7.9   Conclusion

Thus, IoT integrated with the predictive maintenance facilitates the accurate forecasting of failure of a system. Costs associated with the failure and malfunction of system due to improper diagnostics of a system was found to be high and can be avoided through proper maintenance with the aid of IoT. Prescheduling of periodic maintenance also saves the time and huge amount of costs yet to invest in overcoming the unexpected failures. With the help of periodic maintenance, the optimal conditions for prediction of hazardous conditions and their impacts were assessed clearly using IoT. Thus, IoT coupled predictive maintenance in various sectors, such as automotive, health, manufacturing, supply chain, energy management, packaging, food sectors etc., leads to minimal time for maintenance and minimal capital investment of entire system. Some of the advantages of IoT-predictive maintenance were collection and execution of wrong data, higher initial investment, and occurrence of error in collection and assessment of periodically varying data, such as weather, age-related diagnostics, etc.

References

1. Lee, J. *et al.*, New thinking paradigm for maintenance innovation design. *IFAC Proc. Vol.*, 47, 3, 7104–7109, 2014, https://doi.org/10.3182/20140824-6-ZA-1003.02519.
2. Bousdekis, A. *et al.*, Decision making in predictive maintenance: Literature review and research agenda for industry 4.0. *IFAC-PapersOnLine*, 52, 13, 607–612, 2019, https://doi.org/10.1016/j.ifacol.2019.11.226.

3. Xu, X., Chen, T., Minami, M., Intelligent fault prediction system based on Internet of Things. *Comput. Math. Appl.*, 64, 5, 833–839, 2012, https://doi.org/10.1016/j.camwa.2011.12.049.

4. Civerchia, F. *et al.*, Industrial Internet of Things monitoring solution for advanced predictive maintenance applications. *J. Ind. Inf. Integr.*, 7, 4–12, 2017, https://doi.org/10.1016/j.jii.2017.02.003.

5. Parpala, R.C. and Iacob, R., Application of IoT concept on predictive maintenance of industrial equipment, in: *MATEC Web of Conferences*, EDP Sciences, 2017, https://doi.org/10.1051/matecconf/201712102008.

6. Pech, M., Vrchota, J., Bednář, J., Predictive maintenance and intelligent sensors in smart factory. *Sensors*, 21, 4, 1470, 2021, https://doi.org/10.3390/s21041470.

7. Wang, K., Intelligent predictive maintenance (IPdM) system-Industry 4.0 scenario. *WIT Trans. Eng. Sci.*, 113, 259–268, 2016.

8. Zonta, T. *et al.*, Predictive maintenance in the industry 4.0: A systematic literature review. *Comput. Ind. Eng.*, 150, 106889, 2020, https://doi.org/10.1016/j.cie.2020.106889.

9. Vargas-Salgado, C. *et al.*, Low-cost web-based supervisory control and data acquisition system for a microgrid testbed: A case study in design and implementation for academic and research applications. *Heliyon*, 5, 9, e02474, 2019, https://doi.org/10.1016/j.heliyon.2019.e02474.

10. Breivold, H.P. and Sandström, K., Internet of Things for industrial automation–Challenges and technical solutions, in: *2015 IEEE International Conference on Data Science and Data Intensive Systems*, IEEE, 2015, https://doi.org/10.1109/DSDIS.2015.11.

11. Petracca, M. *et al.*, WSN and RFID Integration in the IoT scenario: An advanced safety system for industrial plants. *J. Commun. Software Syst.*, 9, 1, 104–113, 2013, https://doi.org/10.24138/jcomss.v9i1.162.

12. Wang, J. *et al.*, A new paradigm of cloud-based predictive maintenance for intelligent manufacturing. *J. Intell. Manuf.*, 28, 5, 1125–1137, 2017, https://doi.org/10.1007/s10845-015-1066-0.

13. Nguyen, K.T. and Medjaher, K., A new dynamic predictive maintenance framework using deep learning for failure prognostics. *Reliab. Eng. Syst. Saf.*, 188, 251–262, 2019, https://doi.org/10.1016/j.ress.2019.03.018.

14. Selcuk, S., Predictive maintenance, its implementation and latest trends. *Proc. Inst. Mech. Eng., Part B: J. Eng. Manufact.*, 231, 9, 1670–1679, 2017, https://doi.org/10.1177/0954405415601640.

15. Tortorella, G.L., Silva, E., Vargas, D., An empirical analysis of total quality management and total productive maintenance in industry 4.0, in: *Proceedings of the International Conference on Industrial Engineering and Operations Management (IEOM)*, 2018.

16. Bukhsh, Z.A. *et al.*, Predictive maintenance using tree-based classification techniques: A case of railway switches. *Transp. Res. Part C: Emerg. Technol.*, 101, 35–54, 2019, https://doi.org/10.1016/j.trc.2019.02.001.

17. Carvalho, T.P. *et al.*, A systematic literature review of machine learning methods applied to predictive maintenance. *Comput. Ind. Eng.*, 137, 106024, 2019, https://doi.org/10.1016/j.cie.2019.106024.

18. Li, D. *et al.*, Human-centred dissemination of data, information and knowledge in industry 4.0. *Proc. CIRP*, 84, 380–386, 2019, https://doi.org/10.1016/j.procir.2019.04.261.

19. Herrmann, F., The smart factory and its risks. *Systems*, 6, 4, 38, 2018, https://doi.org/10.3390/systems6040038.

20. Lee, S.M., Lee, D., Kim, Y.S., The quality management ecosystem for predictive maintenance in the industry 4.0 era. *Int. J. Qual. Innov.*, 5, 1, 1–11, 2019, https://doi.org/10.1186/s40887-019-0029-5.

21. Farooq, B. *et al.*, Data-driven predictive maintenance approach for spinning cyber-physical production system. *J. Shanghai Jiaotong Univ. (Science)*, 25, 4, 453–462, 2020, https://doi.org/10.1007/s12204-020-2178-z.

22. Dalzochio, J. *et al.*, Machine learning and reasoning for predictive maintenance in Industry 4.0: Current status and challenges. *Comput. Ind.*, 123, 103298, 2020, https://doi.org/10.1016/j.compind.2020.103298.

23. Goodman, D.L., Wood, S., Turner, A., Return-on-Investment (RoI) for electronic prognostics in mil/aero systems, in: *IEEE Autotestcon, 2005*, IEEE, 2005.

24. Wood, S.M. and Goodman, D.L., Return-on-Investment (RoI) for electronic prognostics in high reliability telecom applications, in: *INTELEC 06-Twenty-Eighth International Telecommunications Energy Conference*, IEEE, 2006, https://doi.org/10.1109/INTLEC.2006.251619.

25. Feldman, K., Jazouli, T., Sandborn, P.A., A methodology for determining the return on investment associated with prognostics and health management. *IEEE Trans. Reliab.*, 58, 2, 305–316, 2009, https://doi.org/10.1109/TR.2009.2020133.

26. Chang, M.-H., Pecht, M., Yung, W.K., Return on investment associated with PHM applied to an LED lighting system, in: *2013 IEEE Conference on Prognostics and Health Management (PHM)*, IEEE, 2013, https://doi.org/10.1109/ICPHM.2013.6621434.

27. Wang, W. and Pecht, M., Economic analysis of canary-based prognostics and health management. *IEEE Trans. Ind. Electron.*, 58, 7, 3077–3089, 2010, https://doi.org/10.1109/TIE.2010.2072897.

28. Formica, T. and Pecht, M., Return on investment analysis and simulation of a 9.12 kilowatt (kW) solar photovoltaic system. *Sol. Energy*, 144, 629–634, 2017, https://doi.org/10.1016/j.solener.2017.01.069.

29. Banks, J. and Merenich, J., Cost benefit analysis for asset health management technology, in: *2007 Annual Reliability and Maintainability Symposium*, IEEE, 2007, https://doi.org/10.1109/RAMS.2007.328097.

30. Banks, J. *et al.*, How engineers can conduct cost-benefit analysis for PHM systems. *IEEE Aerosp. Electron. Syst. Mag.*, 24, 3, 22–30, 2009, https://doi.org/10.1109/MAES.2009.4811085.

31. Pozzi, M. *et al.*, A framework for evaluating the impact of structural health monitoring on bridge management, in: *Proc. 5th Int. Conf. Bridge Maintenance, Safety Manage*, 2010, https://doi.org/10.1201/b10430-91.

32. Memarzadeh, M. and Pozzi, M., Value of information in sequential decision making: Component inspection, permanent monitoring and system-level scheduling. *Reliab. Eng. Syst. Saf.*, 154, 137–151, 2016, https://doi.org/10.1016/j.ress.2016.05.014.

33. Raiffa, H. and Schlaifer, R., *Applied statistical decision theory*, Wageningen Univ. & Res., - Netherland, 1961.

34. Haddad, G., Sandborn, P., Pecht, M., Using maintenance options to maximize the benefits of prognostics for wind farms. *Wind Energy*, 17, 5, 775–791, 2014, https://doi.org/10.1002/we.1610.

35. Zio, E. and Compare, M., Evaluating maintenance policies by quantitative modeling and analysis. *Reliab. Eng. Syst. Saf.*, 109, 53–65, 2013, https://doi.org/10.1016/j.ress.2012.08.002.

36. Youn, B.D., Hu, C., Wang, P., Resilience-driven system design of complex engineered systems, *J. Mech. Des.*, 113, 10, 1–15, 2011, https://doi.org/10.1115/DETC2011-48314.

37. Yoon, J.T. *et al.*, A newly formulated resilience measure that considers false alarms. *Reliab. Eng. Syst. Saf.*, 167, 417–427, 2017, https://doi.org/10.1016/j.ress.2017.06.013.

38. Yoon, J.T. *et al.*, Life-cycle maintenance cost analysis framework considering time-dependent false and missed alarms for fault diagnosis. *Reliab. Eng. Syst. Saf.*, 184, 181–192, 2019, https://doi.org/10.1016/j.ress.2018.06.006.

39. Compare, M., Bellani, L., Zio, E., Reliability model of a component equipped with PHM capabilities. *Reliab. Eng. Syst. Saf.*, 168, 4–11, 2017, https://doi.org/10.1016/j.ress.2017.05.024.

40. Compare, M., Bellani, L., Zio, E., Availability model of a PHM-equipped component. *IEEE Trans. Reliab.*, 66, 2, 487–501, 2017, https://doi.org/10.1109/TR.2017.2669400.

41. Saxena, A. *et al.*, Evaluating algorithm performance metrics tailored for prognostics. *2009 IEEE Aerospace Conference*, IEEE, 2009, https://doi.org/10.1109/AERO.2009.4839666.

42. Cannarile, F. *et al.*, An unsupervised clustering method for assessing the degradation state of cutting tools used in the packaging industry, in: *ESREL 2017*, 2017, https://doi.org/10.1201/9781315210469-119.

43. Pipe, K., Practical prognostics for condition based maintenance, in: *2008 International Conference on Prognostics and Health Management*, IEEE, 2008, https://doi.org/10.1109/PHM.2008.4711424.

44. Hess, A. *et al.*, Challenges, issues, and lessons learned chasing the "Big P": Real predictive prognostics part 2, in: *2006 IEEE Aerospace Conference*, IEEE, 2006, https://doi.org/10.1109/AERO.2005.1559666.

45. Rigamonti, M. *et al.*, Ensemble of optimized echo state networks for remaining useful life prediction. *Neurocomputing*, 281, 121–138, 2018, https://doi.org/10.1016/j.neucom.2017.11.062.

46. Winnig, L., GE's big bet on data and analytics. *MIT Sloan Manage. Rev.*, 57, 3, 1–21, 2016.

47. Rahim, M.A. *et al.*, Evolution of IoT-enabled connectivity and applications in automotive industry: A review. *Veh. Commun.*, 27, 100285, 2021, https://doi.org/10.1016/j.vehcom.2020.100285.

48. Wang, K. and Wang, Y., Towards a next generation of manufacturing: Zero-Defect Manufacturing (ZDM) using data mining approaches, in: *Data mining for Zero-Defect Manufacturing*, Tapir Academic Press, New York, 2012, https://doi. org/10.1007/s40436-013-0010-9, https://doi.org/10.1007/s40436-013-0010-9.

49. Wang, K.-S., Towards zero-defect manufacturing (ZDM)-a data mining approach. *Adv. Manuf.*, 1, 1, 62–74, 2013, https://doi.org/10.1007/s40436-013-0010-9.

50. Liu, X. and Sun, Y., Information integration of CPFR in inbound logistics of automotive manufactures based on internet of things. *J. Comput.*, 7, 2, 349–355, 2012, https://doi.org/10.4304/jcp.7.2.349-355.

51. Nasr, E., Kfoury, E., Khoury, D., An IoT approach to vehicle accident detection, reporting, and navigation, in: *2016 IEEE International Multidisciplinary Conference on Engineering Technology (IMCET)*, IEEE, 2016, https://doi.org/10.1109/IMCET.2016.7777457.

52. Handte, M. *et al.*, An internet-of-things enabled connected navigation system for urban bus riders. *IEEE Internet Things J.*, 3, 5, 735–744, 2016, https://doi.org/10.1109/JIOT.2016.2554146.

53. Raj, J.T. and Sankar, J., IoT based smart school bus monitoring and notification system, in: *2017 IEEE Region 10 Humanitarian Technology Conference (R10-HTC)*, IEEE, 2017, https://doi.org/10.1109/R10-HTC.2017.8288913.

54. Javaid, S. *et al.*, Smart traffic management system using Internet of Things, in: *2018 20th International Conference on Advanced Communication Technology (ICACT)*, IEEE, 2018, https://doi.org/10.23919/ICACT.2018.8323770.

55. Pyykönen, P. *et al.*, IoT for intelligent traffic system, in: *2013 IEEE 9th International Conference on Intelligent Computer Communication and Processing (ICCP)*, IEEE, 2013, https://doi.org/10.1109/ICCP.2013.6646104.

56. Sumi, L. and Ranga, V., Intelligent traffic management system for prioritizing emergency vehicles in a smart city. *Int. J. Eng.*, 31, 2, 278–283, 2018, https://doi.org/10.5829/ije.2018.31.02b.11.

57. Muthuramalingam, S. *et al.*, IoT based intelligent transportation system (IoT-ITS) for global perspective: A case study, in: *Internet of Things and Big Data Analytics for Smart Generation*, pp. 279–300, Springer, Switzerland, 2019, https://doi.org/10.1007/978-3-030-04203-5_13.

58. Uddin, M. *et al.*, Design and application of radio frequency identification systems. *Eur. J. Sci. Res.*, 33, 3, 438–453, 2009.

59. Roy, A. *et al.*, Smart traffic & parking management using IoT, in: *2016 IEEE 7th Annual Information Technology, Electronics and Mobile Communication Conference (IEMCON)*, IEEE, 2016, https://doi.org/10.1109/IEMCON.2016.7746331.

60. Krishna, A.V. and Naseera, S., Vehicle detection and categorization for a toll charging system based on tesseract ocr using the IoT, in: *International Conference on Communications and Cyber Physical Engineering 2018*, Springer, 2018.

61. Mukhopadhyay, D. *et al.*, An attempt to develop an IoT based vehicle security system, in: *2018 IEEE International Symposium on Smart Electronic Systems (iSES)(Formerly iNiS)*, IEEE, 2018, https://doi.org/10.1109/iSES.2018.00050.

62. Metan, J. and Patil, K.K., Data acquisition in car using IoT, in: *AIP Conference Proceedings*, AIP Publishing LLC, 2018, https://doi.org/10.1063/1.5078985.

63. Pascual Espada, J., Yager, R.R., Guo, B., Internet of things: Smart things network and communication. *J. Netw. Comput. Appl.*, 42, 118–119, 2014, https://doi.org/10.1016/j.jnca.2014.03.003.

64. Lai, Y.-L., Chou, Y.-H., Chang, L.-C., An intelligent IoT emergency vehicle warning system using RFID and Wi-Fi technologies for emergency medical services. *Technol. Health Care*, 26, 1, 43–55, 2018, https://doi.org/10.3233/THC-171405.

65. Jamil, M.S. *et al.*, Smart environment monitoring system by employing wireless sensor networks on vehicles for pollution free smart cities. *Proc. Eng.*, 107, 480–484, 2015, https://doi.org/10.1016/j.proeng.2015.06.106.

66. Erkuş, U. and Özkan, T., Young male taxi drivers and private car users on driving simulator for their self-reported driving skills and behaviors. *Transp. Res. Part F: Traffic Psychol. Behav.*, 64, 70–83, 2019, https://doi.org/10.1016/j.trf.2019.04.028.

67. Park, S.J. *et al.*, Intelligent in-car health monitoring system for elderly drivers in connected car, in: *Congress of the International Ergonomics Association*, Springer, 2018, https://doi.org/10.1007/978-3-319-96074-6_4.

8

Role of IoT in Product Development

Bhuvanesh Kumar M.[1,2]*, Balaji N. S.[1], Senthil S. M.[2] and Sathiya P.[1]

[1]Department of Production Engineering, National Institute of Technology,
Tiruchirappalli, India
[2]Department of Mechanical Engineering, Kongu Engineering College, Erode, India

Abstract

The increasing product complexity emerges digitalization of product manufacturers to meet the challenges, such as uncertainties with technological and business settings. The customers are looking for innovative solutions to understand better the product features rather than the complexity associated with its functionality. From the manufacturers' perspective, developing complex and smart products requires huge amount of data to be collected, assessed, processed, and enhanced based on the usage and requirement. Internet of Things (IoT) will be the most beneficial technology when integrated in every phase of design and development of a product. The IoT, when integrated with the product development become a comprehensive and a practical approach. This chapter aims to present the transformation of conventional manufacturing to modern manufacturing called Industry 4.0, and the insights on how IoT integration helps the manufacturers to digitalize the development and production activities. Also, the challenges related to the implementation of IoT in product development phases are discussed. This chapter will provide the needs, key learning, and guidance to the product development people and practicing managers to know how IoT can be linked to the product development activities.

Keywords: Product development, architecture, Industry 4.0, Internet of Things (IoT), modern manufacturing, digitalization

*Corresponding author: bhuvanesh85@gmail.com

R. Rajasekar, C. Moganapriya, P. Sathish Kumar and M. Harikrishna Kumar (eds.) Integration of Mechanical and Manufacturing Engineering with IoT: A Digital Transformation, (215–234) © 2023 Scrivener Publishing LLC

8.1 Introduction

New products drive business. The industries are always looking for different methodologies and concepts to develop their products. To meet this requirement, enterprises opt for new product development to sustain themselves in the market competition. Product development is the process of step-by-step activities, from the perception of market opportunities to the end of product manufacturing, sales, and delivery. The product development begins with formulating customer needs, concept generation, prototype modeling, and testing specification, matching customer demand. According to the required product design scope, the specification is developed and optimized. This chapter aims to provide an overview to understand the product development process while concentrating on the design aspects, manufacturing aspects, marketing strategy, and the organization's objectives.

On the other hand, to shorten the time for developing a new product, industries generally adapt redesign concepts. The redesign of a product starts from the acquisition of an existing product, product modification, and improvements from the own research developments. In research and development, some descriptive and standard design methods have been developed and used for modeling and addressing engineering design issues [1–6]. However, there are few ways to focus on the redesign problem (adaptation, variation, etc.) [7]. The steps in the actual design process include the collection of customer requirements, planning and development specifications, idea generation concepts, product embodiment, development of rapid models and testing, manufacturing design process, benchmark process, but redesign process pay attention to an extra step called reverse engineering [8]. Reverse engineering positively affects the design process to match the product with its intended industrial applications [9].

The inventions, technological advancements, energy sources, and methods alone cannot report the industrial revolution. Still, these factors were the most important in the competitive economy for the last couple of centuries. The industrial revolution that happened in the earlier periods is depicted in Figure 8.1. From the continuation of the product development process, Industry 3.0 for automation is already exists, and now the industry and designer are entirely focusing on new product development based on Industry 4.0 concepts. Presently, Internet of Things (IoT) based concepts are very emerging technologies in Industry 4.0. Many definitions exist for IoT around the world since a unique definition has not been created yet.

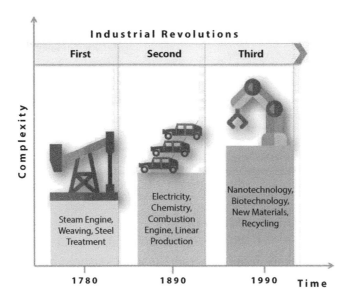

Figure 8.1 Industrial revolutions happened in earlier decades [10].

The IoT is the interconnection of distinctively identifiable networks, usually equipped with ubiquitous intelligence. The IoT emerges everywhere due to its processability of many data by integrating each object through embedded systems. This leads to a high degree of a distributed network that enables communication among many devices [11]. Since product development involves data collected from different steps, the integration of IoT can handle the product development process effectively through a connected network.

8.1.1 Industry 4.0

The primary ideas of industry 4.0 were first posted in the year 2012 by Henning Kagermann. He constructed the strategy platform for industry 4.0 and submitted it to the Academy of Science and Engineering in Germany. In the recent era, industry 4.0 evolved as an advanced state of the industrial revolution since it can organize and control the value chain of a product for its complete lifecycles. Intelligent innovations, cloud-based manufacturing, intelligent solutions, and a clever delivery chain are four primary drivers of Industry 4.0. The other drivers are given in Figure 8.2. The following points state the conditions and goals of industry automation 4.0 from the literature [12, 13]:

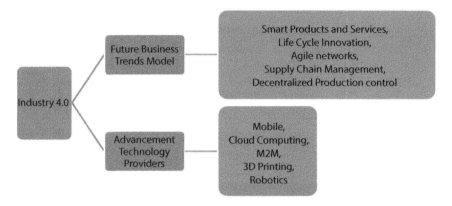

Figure 8.2 Primary drivers of Industry 4.0.

1. Manufacture desires to accept the small, medium, and excessive entities with the aid of modifying the product category.
2. Self-reputation of components and tracking merchandise are done by machines.
3. Human-machine interface is better.
4. Optimization of manufacturing process primarily based on the information of the net gear factors.
5. Radical alternate inside the commercial enterprise version supports modifying the styles of interplay with the fee chain.

The cyber-physical systems (CPS) are the crucial component in implementing Industry 4.0. The CPS mainly provide new functions and capabilities based on network connectivity with machine and system. While innovative products provide the platform for business development, new technologies enable the digitalization of delivered products. This helps in achieving the process capabilities of CPS. These smart-systems come from long-term innovation interconnected with the product life cycle. Escalating innovation has two information flows from inside to outside and outside to inside. The intelligence of advanced product life cycle control (PLC) structure forms a linked modernized life cycle. Thus, smart products can be operated through mobile phones using applications from anywhere, any location.

A smart supply chain is a combination of connected supply chains and agile collaboration networks (ACN). The ACN designates the horizontal integration trend that allows focusing their core competencies of manufacturers while offering the products in the customized market. Vertical supply networks connect supply chains, allowing the integration of physical processes with automation. Finally, the factory having the components of decentralized production control and data-driven operational excellence are called the smart factory. The machines are sequenced on a network in self-organization or companies, and their process configuration allows the decentralized production control. The rich data generated during the manufacturing process serves as a stable basis for this data-driven operational excellence [14].

The architecture of a mobile phone contains internet technologies, and other features, such as WiFi facilities, a cellular connection, and other similar wireless communication technologies are also included in current trend products. Due to this advancement, a large amount of data already available in a constant location can now be used anytime, anywhere. As far as Industry 4.0 is concerned, the WiFi or mobile internet data is essential for connecting the manufacturing environments, such as timeline real data capture, object tagging, accessibility of the data, and network to machine or module communication. The CPS generates large amounts of data that must be stored and processed so that it can be available as requested by the application. And the results of any type of analysis must be available anywhere in the world, at any time. There is another term called "cloud computing" technologies, which refers to the applications, platforms, and organization resolutions that offer resources, data on-demand basis per usage to the users of common community networks. As an essential part of Industry 4.0, cloud computing technology supports this boundless data flow. Data analysis aims to generate business insights from the data pool by categorizing designs and dependencies. The quantity of data available to manufacturers will expand as the number of CPS in the manufacturing industry and smart goods while the market grows. However, this precious source is often ignored. Manufacturers will be able to examine their operational processes and company performance with the aid of data analysis, uncover and explain inefficiencies, and even anticipate future occurrences.

Machine-to-Machine communication (M2M) is at the core of Industry 4.0's impact on the workshop. M2M refers to a technology that allows the automatic exchange of information between the CPS that make up the production environment. M2M can be regarded as an essential technology of the IoT. The entire production workshop can use sophisticated integrated sensors and actuators technology to transmit meaningful information and form an interface between the real world and the virtual world. Customer groups are enterprise-level products that rely on employee cooperation to provide more dynamic and content-rich interactions with customer suppliers, distributors, and collaborators. Community platform components may now be found in a wide range of different corporate applications.

This technology has dramatically promoted and improved the relationship between people in the networked industrial environment.

3D printing technology, generally known as additive manufacturing, produces the real 3D prototypes models directly from the virtual data and codes. Businesses use this new generation technology less often due to costly manufacturing quotes, availability of materials in the required form [15, 16]. As current improvements lighten those weaknesses, possibilities are plentiful to use additive manufacturing frequently to work with customized product requirements. IoT integration permits rapid prototyping from a decentralized production system to print the product closest to the customer requirements through applications and websites, thus eliminating difficulties in handling intermediate manufacturing procedures, warehousing, and logistics. In the past few decades, robotics innovation has greatly enhanced the automation process, and robots can be used almost in every production process. Particularly, sensors-based products and gadget vision coupled with progressed synthetic intelligence permit advanced robots to meet their position in production as unbiased efficient units safely along with shop ground employees. In Industry 4.0, this robotics technology will be significant for system performance and lowering complexity [17].

Rapidly increasing customer needs, the speedy development of digital transformation and the growth in accessing the Internet are moving the market tendencies towards smart products, linked the IoT products. This scenario forces the industry to design, expand the variants, and mass-production for customized smart products at affordable charges to maintain a leading position in the competition. Within the early levels of product development phases, the entire life cycle of a product should be considered for new or redesign. Here, concepts that include design engineering, mass customization, flexible manufacturing planning, and intelligent manufacturing systems integrated with customers can deal with new demanding situations and alternate the manner products are treated in their life cycle. The idea of this chapter is to describe the importance of a scientific product improvement technique for the development of customizable smart products. This entails more effective approaches than industrially produced mechatronics products. According to this prospect, IoT concepts in the product development process are discussed in this chapter.

8.2   Need to Understand the Product Architecture

An industry that uses the potential of digitalization to increase productivity is classified as Industry 4.0, and it is considered a new revolution in modern industries. Enterprises are gradually accepting this revolution and progressively adopting digitalization, information, and communication technologies. It is expected that the industry will be digitally integrated in various ways in the future; they are called digital factories, smart factories, or information factories [18]. Industry 4.0 is characterized by supporting the incorporation and virtualized use of product design process and production planning information, and the Internet creates smart products.

Understanding how an organization manages the knowledge related to the product architecture it designs is a crucial challenge for companies developing new smart products. As emphasized by Henderson and Clark [17], "Architectural knowledge tends to be embedded in the structure and information processing procedures of established organizations". Therefore, the company design teams dealing with the new architecture must know how they accomplish the embedded knowledge of the developed products. Since the process involves a large number of design participants and physical components, this is particularly important in complex product development. Unfortunately, there are few methods and tools available to meet this challenge.

Design thinking is a broad, user-centered approach that employs a systematic use of observation, questioning, brainstorming, and other adjustment techniques at various phases of a multi-iterative process. The product design steps are introduced in a flow chart, as shown in Figure 8.3. The actual implementation stage develops the idea into a marketable product/service before proceeding.

Products are gradually evolved into an advanced combination of heterogeneous equipment and mechanical and mechatronics technology, resulting in complex edges between these fields. This mechatronics product is further integrated into the network for communication. They are generating, exchanging, and using data to complete future functions. Such products are classified as IoT devices. The general representation of the elements of an IoT device is shown in Figure 8.4.

IoT device customization might take the shape of hardware, software, communications, Internet/cloud platforms, or end-user services/applications. When personal IoT devices need professional work situations that limit WiFi networks and users do not want to utilize hotspot phones, communication methods like Bluetooth or Sigfox may be the best option. Customers can only get appropriate functions from the firm, determined not just by limits but also by needs. For example, when a client does not want to utilize all of the device capabilities and does not want to pay for those that are not used. This improves the customer experience with the product, allowing it to suit particular demands while avoiding misunderstanding and resource and energy waste [19, 20].

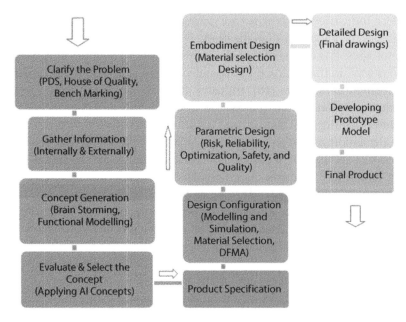

Figure 8.3 Product design steps.

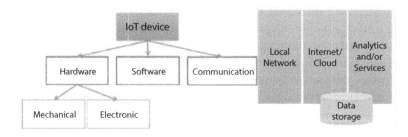

Figure 8.4 General elements of IoT devices [19].

8.3 Product Development Process

Creating a new product design is the product development, which is sold by a company or a company to its end users. Development is mutually referred to as the whole process of classifying market opportunities, creating products to attract the recognized market, modifying, improving the product and finally testing it until it is ready for production.

In the primary stages of the product development process, the designers identify and analyze the various risk factors. So, the product development process can be understood as a risk management system. As the process

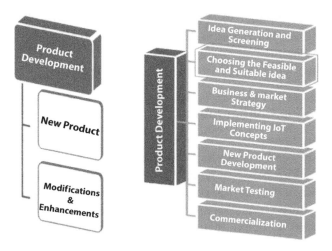

Figure 8.5 New product development process steps.

progresses, risk reduction is the key to reject the uncertainty, and to verify the product's function. After completing the process, the design team should be confident about the working nature of the product and market expectations of the product. The new product development process steps are shown in Figure 8.5. Many models and process methods were adopted to understand, support and improve the design and development of physical products and software process-based products. These methods or models define the project's macro-level structure, end-to-end task flow, meso-level or individual process steps and their direct micro-level background [21].

8.3.1 Criteria to Classify the New Products

The new products are classified according to the following criteria,

- a) New to the world market (Ex. Electric Vehicles)
- b) New product Line (Ex. Launching new configuration mobiles, cars)
- c) In addition to the existing lines (Ex. Jio launches Android Mobile phone)
- d) Improvements and revision to the current Product (Ex. Nokia enter to Android)
- e) Cost reductions (Ex. Production Cost)
- f) Repositions (Ex. Pharmachetuicals)

The designer should collect the data based on the above classifications; then, information is grouped and optimized for further process. The above product development steps are synchronized with IoT devices, and then the automated system will start the manufacturing process.

8.3.2 Product Configuration

It is a process that constitutes processing of a product according to the person's wishes called a product configurator. It offers the possible product configurations based on complete consumer evaluation. To develop a product configurator, feasible confirmations are required to define the process. The Feature-Orientated Domain Evaluation (FODA) model defines the exceptional design configurations in one manner. Obbink and Pohl [22] have explored the core relationships between the various process features. That is, in turn, acts because the entry for the software program, software team to expand a product configurator. The configurator is no longer the most effective collects statistics from the purchaser. However, it also plays a role that consists of the records flow into and out of the employer. The statistics glide, which ensues about a product function configurator, is provided in Figure 8.6.

The selected configuration through the client via the device affords an entire sequence that the enterprise can make a complete picture to manufacture the product. The inner structure similarly can assist the investor knowledge with the aid of supplying the visualization of the product with its capabilities and possible transport time. The configurator likewise facilitates details about the product and the subsequent step of manufacturing, making plans through product information management gadgets. All of

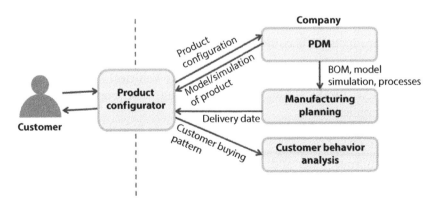

Figure 8.6 Product configuration functions.

the editions can be accompanied by their production strategies and assembly procedures. The records and data will assist in production planning and logistics, production and assembly configurations [23]. These steps are performed similar to the way suggested by Trentin, Perin [24]. The product configurator has a good effect on the overall performance of the factory's time. Using the product configurator is likely to reduce dependence on temporary solutions to meet the wishes of different customers, reduce product configuration errors, and reduce designers and process engineers. The workload is itself a configuration process. Therefore, it is strongly recommended to broaden the product configurator because it can help customers configure products and efficiently develop and produce versions.

8.3.3   Challenges in Product Development while Developing IoT Products (Data-Driven Product Development)

The system product development methodology drives the product development process. However, the existing methods lack special attention to customizable IoT equipment development. In the product development process, it is facing a lot of challenges for focusing on the current research and development. In this section, the discussion is focused on (a) The challenges in customizable IoT device, and (b) Fit-gap analysis. However, at the time of the development process, there are challenges to be faced such as the answer to the questions related to customization was whichever not available, or the method was insufficient. This has led to a gap in the development of customizable IoT method equipment. At the initiation time of the product development process, the requirements list was at critical stage. It is most important to consider all the related demand requirement lists, which will later be used as a guide for developing products. Therefore, it is subject to the required category defined by Beitz, Pahl [25].

In the new product development and design process, the following challenges are faced in the IoT devices and common challenges are listed below [26–29].

- Maintain the data privacy and security
- Requirement of vast storage capacity and analysis platform for data
- Power consumption and availability
- Challenge due to Chaos (device failure in the IoT network)
- Providing communication to multiple-devices (running capacity)

- Integrating customers
- Reduce development costs
- Product transportation and delivery

The challenges create the opening for systematized product development of new IoT products. Basic categories of fit gaps based on challenges develop analysis to determine suitable products development methods. These models and methods develop application-based IoT devices.

To confront the challenges found in the new product development process, the following criteria were identified to fit the gap:

- Engineering requirements and management needs
- Product life and Reliability considerations
- Variation in management
- Product configurator
- Make or purchase decision
- Concurrent engineering
- Utilization of information management system
- IoT service considerations and facilities

The three levels of standards meeting criteria can be fixed: covered, uncovered, and partially covered. The basic knowledge of the standard is very detailed and incorporates methods or models in the systems are called "Covered". The "Partly Covered" means only partial information standards are mentioned, but relevant information is missing to the technique or realization. They are not overwritten as a name suggest, also with no mention of standard methodology or model. Hence, based on the above methods, the product team can gather more information and forward it to the IoT sector to develop the automated machine for manufacturing the products.

8.3.4 Role of IoT in Product Development for Industrial Applications

By 2025, It is expected that there will be hundreds of connected gadgets and billions of connected gadgets via IoT devices [30]. If this becomes a reality, the company should design the IoT-ready products and must manufacture related products. The development process of the IoT is more effective, united, and streamlined. Now, it is necessary to complete the value of IoT readiness.

Product's value is mainly based on their features and functions. The IoT-ready products' value will also be reflected in the ability to aggregate its knowledge content and integrate vertically and horizontally in the cloud within the future world [31]. The product development process (PDP) for IoT can be well-designed to realize the entire IoT to a large extent [32, 33]. The evaluation of the manufacturing sector is illustrated in Figure 8.7. The development transformed the manufacturing sector from manual operations to IoT-implemented advanced techniques. This includes one or more automated functions, from supplementing human operators to replacing human operators. Despite of elimination of human workers, industries committed strongly for digital transformation. According to the analysis reported by Cline [34], 27% of the manufacturer around the world have already planned to achieve the so called "industrial IoT" in every aspect of the product development and manufacturing. From consumer side, the transformation arises from the small electronic gadgets, high-level connectivity appliances such as laptop computers, tablets, and smart phones. This can be extended to the applications like work place, living space, and transport that are integrated with sophisticated electronics.

From the manufacturer's side, the transformation empowers them to innovatively develop, and manufacture products with endless connectivity resources. This evolution is a subset of IoT called industrial IoT, which

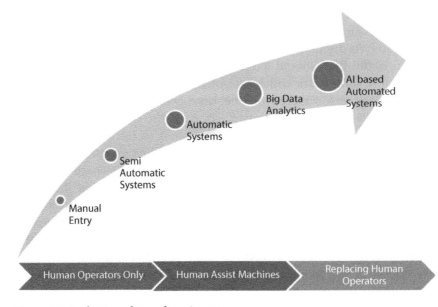

Figure 8.7 Evaluation of manufacturing sector.

connects the product development with the manufacturing activities. It is not limited to the development of product but also extendable to service activities. The IoT-enabled processes not only cost-effective but also helps the organizations to enhance the uptime and availability of process equipment. Needs for quick release of products, progressive innovative features, widening market opportunities push the manufacturers to stay in the global competition with the aid of cutting-edge technologies for product development. According to the report by Aberdeen Group, 2017, 48% of the manufacturers realize peak pressure while developing products to meet high quality, high performance requirements of products. Close to the same count, 47% of the manufacturer reported that the need for very complex but customized product is the main reason for their interest to transform.

The pressures for manufacturers in product development include,

(a) Demand to release the products as quickly as possible before competitors
(b) Demand from market for higher performance, and quality products
(c) Demand from market for more customized yet complex products
(d) Requirements to launch featured products at low cost
(e) Sticking in accordance with standards and regulations

These pressures make them to think in rational way to design products through IoT. As discussed, the IoT-based products possess more complexity. Hence, the PDP should lodge higher performance, reliability, more functionality, form factor without negotiating the power.

The present era centered on the latest-information, a communication approach called "digital thread" connects the isolated departments/units to the manufacturing processes. This offers an integrated view of its advantages throughout the entire life cycle of product manufacturing. The industries falls under best-in-class category finds this approach as greatly compelling. Those industries adopt digital thread throughout the entire product life cycle 2.4 times higher than other industries. The transformation drives the modern product design and development, manufacturing, and services. In addition, the complete digital thread integration is the key that enables the manufacturers to attain success at inception, shorten the cycle time, and to effectively manage the complexity associated with today's products.

8.3.5   Impacts and Future Perspectives of IoT
in Product Development

Despite of the risks and challenges portrayed in previous sections, there is a huge potential for IoT-enabled industrial activities and the product life cycles. Golovatchev, Chatterjee [35] have reported exemplary cases from three different industrial sectors such as automotive, health care, and energy to assess and overcome the risks and challenges.

Automotive sector
As described in earlier sections, IoT will have a huge impact on the smart factories in forthcoming era. We take an example of Intel's Edison wearable's development module. The module incorporates the technology into the usual motions of operator's hands. The module integrates Radio Frequency Identification (RFID) technology with sensors, automatic scanning and motion tracking [36]. It is a best case to understand how IoT can be effectively integrated into the product to be best used in the field of manufacturing engineering and logistics. The technology adoption can assist the operators/workers to perform their job more effectively [37]. The product not only helps the workers with the operating instructions, essential information about the products and processes but also collects the data related to process documentation, and sequence of the job work and transmits to the central storage. The technology-enabled tools works with cleverness to help the worker avoid mistakes and to follow the right sequence of work activities.

As this technology powered application integrated to an information technology (IT) system of automotive original equipment manufacturers (OEM), the data associated with production and handling can be acquired and stored. While the use of IoT-enabled devices from different stages of product manufacturing from the shop floor and the vendors, the integrated system become more powerful in the IT landscape. These devices can communicate each other, and the associated data stored can be assessed and analyzed by the designers for enhancing products during the product development stage. Hence, the development time can be compressed to short spans that will aid to launch the enhanced products soon in the market before the competitor does.

Conversely, the challenge related to the integration with product manufacturing is to align all the IoT devices of every stage with respect to interfaces, software, and updates to timely tie up the technology into actions. This has to be holistically viewed on four functional design realms of product life cycle management (PLM). The PLM design domains include strategy, process, product and IT architectures. The need for integrating IoT in PLM of OEM products is that, IoT can handle product development cycles, product up gradation and updates according to the frequently changing requirements. The major challenge with ProGlove is managing the interactions between the vendors and customers with the perspectives of product enhancements. But different usages statistics from different customers help the designers to upgrade the product more suitable to all the users but quicker analyzing of data is a challenging task to the designers.

Healthcare Sector
Not many but few challenges encountered in the application of IoT in Health and Pharma industry. It costs roughly about 55 million dollars to develop and grant a new drug but despite of huge numbers of drugs being sold, a lot of them are not utilized and simply thrown by consumers. Application of IoT in the form of RFID tags to the drug packages found its own way to move a step ahead in enhancing the drug inventory and counterfeiting. A vast opportunity can be availed to the retailers and pharmacies to store and use the data. It also offers the essential information to the consumers.

Presently growing technology offers the present era to embed the IoT devices to eatable medications. This revolutionary technology advancement enables the medical practitioners to collect data related to health issues, medication brigades, and miss-use of drugs. It will be highly beneficial to the drug manufacturers for testing the drug in case of side effects, thus shortens the product development time. The application of IoT to pharma sector can assist in customizing the products specific to patient need and services. And the technology can also help to identify the drug black markets [35]. Health insurance companies will seriously view these illegal practices with the help of this groundbreaking technology.

The scenario yields greatest advantage to the medicine manufacturers by connecting everyone on the supply chain. The collection of endless data would help them through analytics and advances the product development cycles hence makes a great impact on the costs associated with research and development (R&D). The key learning from data analytics help the sector to shorten the lead time of development and launch medical products into the markets. This is can be a great advantage to meet pandemic situations in health care industry.

Energy and utility sector
To cope up with this digital world, it is essential for a company to cope up with their transition towards digitalization. It is applicable to the energy

sector also, but the challenges being the energy regeneration and digitalization. Overcoming these challenges, the energy sector can build a new successful business models. Since the industries within this sector is facing downward trend in making profits, incorporation of digitalization can yield better ability to be competitive. The media, retail and telecommunication organizations such as Amazon, Google, and Skype were first affected by this transition pressure hence came up with innovative internet usage for their business models.

The transformation towards digitalization forces these organizations to manufacture smart energy products. Due to dynamic and very aggressive market conditions, those industries would require innovative solutions in providing customized smart products along with service provision from their mass production environment. IoT-enabled devices have proven its ability to provide innovative solutions to the smart grid. These devices with their intelligence can convert conventional to data-driven grids, which are smart. This can create many opportunities for both existing and new business organizations like utilities and network providers to enable digitalization.

Smart grids can be highly beneficial to the household utilities. There exist the smart metering devices already which made it possible to measure and monitor the energy consumption and control them remotely which can improvise the energy consumption efficiency. For customers, the technology aid them to see the transparency in billing for energy consumption make them effectively use the energy supply. Implementation as a whole could change the energy pricing policies and will have an impact on the resource utilization.

As a result of radically growing and dynamically varying demand for customized products, the organizations will now require providing more value added services to these applications. Though the smart devices can meet the requirements satisfactorily, the challenge lies with the development of these complex products and services. It is also require incorporating innovative solutions in the design and development stage itself for the present need, again IoT will be the inevitable solution to the product development challenges as they can assess the product portfolio and can make changes swiftly.

8.4   Conclusion

This chapter provides insights into the industrial revolution and transition toward digitalization with the application of IoT. With the rapidly growing technological advancements and its impact on the product development and services, IoT is emerging as a key information and communication technology to manage the entire product development life cycle. Since IoT enable the easy accessibility to the data in real-time usage, manufacturer can monitor, control and quickly change the product specifications and features based on the requirements. This scenario enables the product development phase easier to confront the complexity associated with the product and make its quick release into market, and it also makes changes in the product development architecture to improve the product quality in the early phase of product development.

A developer and practicing manager can gain inputs from the key discussions related to smart factory and how smart devices help consumers to use the products appropriately and effectively. Additionally, new scope for research in the area of IoT implementation not discussed in literature could be realized by understating the future use cases discussed in automotive, energy and utility, and healthcare sectors. The primary problems connected to product complexity, and digitalizing business environments could raise the uncertainties and the managers need to focus specifically on the innovative solutions. Hence, the discussions from this chapter provide the practicing managers guidance on how IoT can be effectively integrated into the projects to face the challenges of different levels. Preferably, the managers could gain the insights on how the IoT integrated systems saves cost, development time, help solving wide range of usage issues, and its competitive advantages.

References

1. Pahl, G. and Beitz, W.J.L., *Engineering design, the design council*, vol. 12, pp. 221–226, Springer, London 1984.
2. Pugh, S., *Total design: Integrated methods for successful product engineering*, Addison-Wesley, Michigan, United States, 1991.
3. Eppinger, S. and Ulrich, K., *Product design and development*, McGraw-Hill Higher Education, New York, 2015.
4. Moehrle, M.G.J.R. and Management, D., How combinations of TRIZ tools are used in companies–results of a cluster analysis. *R&D Management*, 35, 3, 285–296, 2005.
5. Finger, S. and Dixon, J.R., A review of research in mechanical engineering design. Part I: Descriptive, prescriptive, and computer-based models of design processes. *Res. Eng. Des.*, 1, 1, 51–67, 1989.
6. Dym, C.L., *Engineering design: A synthesis of views*, Cambridge University Press, New York, 1994.

7. Sferro, P.R., Bolling, G.F., Crawford, R.H.J.M.E., It's time for the omni-engineer. *Manuf. Eng.*, 110, 6, 60–63, 1993.

8. Ingle, K.A., *Reverse engineering*, McGraw-Hill Professional Publishing, New York, 1994.

9. Otto, K.N. and Wood, K.L., A reverse engineering and redesign methodology for product evolution, in: *International Design Engineering Technical Conferences and Computers and Information in Engineering Conference*, American Society of Mechanical Engineers, 1996.

10. Ortiz, J.H., *Industry 4.0: Current status and future trends*, IntechOpen, London, 2020.

11. Xia, F. *et al.*, Internet Things, . *Int. J. Commun. Syst.*, 25, 9, 1101, 2012.

12. Kagermann, H., Lukas, W.-D., Wahlster, W.J.V.n., Industrie 4.0: Mit dem Internet der Dinge auf dem Weg zur 4. *Ind. Revolution*, 13, 1, 2–3, 2011.

13. Shafiq, S.I. *et al.*, Virtual engineering object/virtual engineering process: A specialized form of cyber physical system for Industrie 4.0. *Procedia Comput. Sci.*, 60, 1146–1155, 2015.

14. Bechtold, J. *et al.*, Industry 4.0-the capgemini consulting view. *Intelligent Industry*, 31, 32–33, 2014.

15. Bhuvanesh Kumar, M. and Sathiya, P., Methods and materials for additive manufacturing: A critical review on advancements and challenges. *Thin-Walled Structures*, 159, 107228, 2020.

16. S.S.R., Yogasundar, S.T., Tamilarasu, S., Varatharaj Kannan, S., Bhuvanesh Kumar, M., Rapid prototyping of human implants with case study. *Int. J. Innov. Res. Sci. Eng. Technol.*, 3, 2, 319–324, 2014.

17. Henderson, R.M. and Clark, K.B., Architectural innovation: The reconfiguration of existing product technologies and the failure of established firms. *Adm. Sci. Q.*, 35, 9–30, 1990.

18. Stark, R., Damerau, T., Lindow, K., Industrie 4.0—Digital redesign of product creation and production in berlin as an industrial location, in: *The Internet of Things*, pp. 171–186, Springer, Berlin, 2018.

19. Gogineni, S.K., Riedelsheimer, T., Stark, R., Systematic product development methodology for customizable IoT devices. *Proc. CIRP*, 84, 393–399, 2019.

20. Thomke, S. and Reinertsen, D., Six myths of product development. *Harv. Bus. Rev.*, 90, 5, 84–94, 2012.

21. Wynn, D.C. and Clarkson, P.J.J.R.i.E.D., Process models in design and development. *Res. Eng. Des.*, 29, 2, 161–202, 2018.

22. Obbink, H. and Pohl, K., *Software product lines: 9th International Conference, SPLC 2005, Rennes, France, September 26-29, 2005, Proceedings*, vol. 3714, Springer, 2005.

23. Forza, C. and Salvador, F., *Product information management for mass customization: Connecting customer, front-office and back-office for fast and efficient customization*, Springer, London, 2006.

24. Trentin, A., Perin, E., Forza, C.J.C.i.I., Overcoming the customization-responsiveness squeeze by using product configurators: Beyond anecdotal evidence. *Comput. Ind.*, 62, 3, 260–268, 2011.

25. Beitz, W., Pahl, G., Grote, K.J.M.B., *Engineering design: A systematic approach*, vol. 71, pp. 125–141, Springer, London, 1996.

26. Hu, S. *et al.*, Product variety and manufacturing complexity in assembly systems and supply chains. *CIRP Annals*, 57, 1, 45–48, 2008.

27. Arshad, R. *et al.*, Green IoT: An investigation on energy saving practices for 2020 and beyond. *IEEE Access*, 5, 15667–15681, 2017.

28. Ling, Z. *et al.*, An end-to-end view of iot security and privacy, in: *GLOBECOM 2017-2017 IEEE Global Communications Conference*, IEEE, 2017.

29. Zorzi, M. *et al.*, From today's intranet of things to a future Internet of Things: A wireless- and mobility-related view. *IEEE Wireless Commun.*, 17, 6, 44–51, 2010.

30. Hayes, J., Research report: IoT features in new product development. Engineering.com. pp. 1–25, 2017. Available at: https://www.engineering.com/ResourceMain.aspx?resid=712

31. Prasad, B., *Product development process for IoT-ready products*, SAGE Publications Sage UK, London, England, 2020.

32. Prasad, B., Lean, integrated & connected framework for developing smart products, in: *Beyond the Internet of Things: Everything Interconnected*, pp. 1–25, Springer, Berlin, 2016.

33. Long, N.J.O., Artificial intelligent (AI) and the future of supply chain. *Opulence*, 2, 1, 1–42, 2018.

34. Cline, G., *Product development in the era of IoT: Tying the digital thread*, Aberdeen Group, Waltham, MA, 2017.

35. Golovatchev, J. *et al.*, The impact of the IoT on product development and management, in: *ISPIM Conference Proceedings*, The International Society for Professional Innovation Management (ISPIM), 2016.

36. Budde, O. and Golovatchev, J., Produkte des intelligenten Markts, in: *Smart Market*, pp. 593–620, Springer, Wiesbaden, 2014.

37. Budde, O. and Golovatchev, J., Descriptive service product architecture for communication service provider, in: *Functional Thinking for Value Creation*, pp. 213–218, Springer, Berlin, 2011.

9

Benefits of IoT in Automated Systems

Adithya K.* and Girimurugan R.

Department of Mechanical Engineering, Nandha College of Technology, Perundurai, Tamil Nadu, India

Abstract

As automation technologies advance, life becomes more straightforward and convenient to use. Due to fast advancements in the realm of automation, human existence is becoming more advanced and superior in every way. The Internet has been ingrained in daily life as its user base has exploded in the last decade, and the Internet of Things (IoT) is the most cutting-edge and rapidly growing Internet technology. Internet of things like sensors and actuators, computing devices, mechanical gear, and people are all connected via the Internet, which is quickly becoming a major part of the internet. The IoT is a critical component of human life because it is a rapidly developing network of ordinary devices ranging from industrial robots to consumer electronics that can share data and perform tasks automatically. A smart city leverages data and technology to improve efficiency and the quality of life for its residents. Determinants of quality of life, such as living expenses, security, time management, job opportunities, and connectivity, have the potential to improve by 10% to 30% as a result of smart city technologies. Smart cities can have advantage from the IoT and smart technology in the following zones: home automation, healthcare automation, manufacturing automation, industrial automation, pollution monitoring, and irrigation automation.

Keywords: IoT, automated system, applications, benefits

9.1 Introduction

Recent years have seen a lot of media attention paid to the advantages, breakthroughs, and improvements of the Internet of Things. The idea of

Corresponding author: adithya6776@gmail.com

R. Rajasekar, C. Moganapriya, P. Sathish Kumar and M. Harikrishna Kumar (eds.) Integration of Mechanical and Manufacturing Engineering with IoT: A Digital Transformation, (235–270) © 2023 Scrivener Publishing LLC

millions of different appliances automatically collaborating over a global network has been around for around three decades, but it has only just begun to expand at breakneck speed. The Internet of Things (also known as IoT) offers countless indisputable advantages, even though some designers take this notion to its logical conclusion by proposing to connect every oven, toothbrush, and garbage can to the Internet [1].

The IoT can be characterized as "objects with identities and virtual personalities in intelligent spaces that link and communicate with one another in societal, medicinal, ecological, and user contexts." In order to provide a wide range of services, vast amounts of money are being invested in the IoT arena. IoT applications are currently being researched in a wide range of social and economic contexts.

9.2 Benefits of Automation

Due to the automation and connectivity characteristics of the Internet of Things, different technologies must be used to provide automatic data transmission, evaluation, and reaction between multiple devices. Automation, for example, is difficult without the use of AI Technology, Cloud Computing, and Computer Vision, while cloud computing and wireless communication technologies significantly simplify connectivity.

9.2.1 Improved Productivity

Simple chores may be robotic with the use of IoT skills, freeing up human resources for more difficult jobs that demand personal talents, particularly creative thinking. In this method, the number of employees can be lowered, resulting in lower business operating costs.

9.2.2 Efficient Operation Management

Furthermore, automatic command over a wide range of operational aspects is one of the primary advantages of smart device connectivity, including stock management, shipment tracking, and the management of fuel and replacement parts. This strategy entails tracking the position of equipment and items through the use of RFID tags and a related network of sensors.

9.2.3   Better Use of Resources

Networked sensors offer automated scheduling and monitoring, which improves access control and water usage. Large and small businesses alike can save a considerable amount of money on water and electricity by installing simple motion detectors. This will increase productivity while also being good for the environment.

9.2.4   Cost-Effective Operation

Owing to the decreased idle stages caused by autonomously controlled and managed conservation, raw material supply, and other manufacturing demands, the system may operate at a higher rate, resulting in increased earnings.

9.2.5   Improved Work Safety

Periodic maintenance is essential for ensuring operating safety and compliance with required standards, in addition to the previously mentioned advantages. On the other hand, safe working conditions improve the company's attractiveness to investors, partners, and employees, improving the reputation and trust of the brand. As a result, smart gadgets help to improve safety by reducing the chance of human error at various stages of business operations. Security may be improved in an organization by using an Internet of Things network such as cameras, motion detectors, and other monitoring devices to stop thefts and even corporate espionage from taking place.

9.2.6   Software Bots

Checking each operation of software or industrial machinery is complex and time consuming for an engineer. Thus, technology bots can be used to detect flaws and notify the engineer, allowing them to rapidly resolve the issue and restart the machinery.

9.2.7   Enhanced Public Sector Operations

The Internet of Things in automation will benefit governments and service-related sectors in a variety of ways. Internet of Things applications can support public sector services in managing traffic, public safety, resource management, and city government.

9.2.8 Healthcare Benefits

Automation in IoT can assist doctors, healthcare experts, and medical employees in detecting and monitoring hazardous conditions. Additionally, it can assist the hospital administration department in managing available beds, ambulances, and other resources for patients [2].

9.3 Smart City Automation

Smart and innovative solutions are becoming increasingly important as cities expand and develop in size in order to increase productivity, improve operational efficiency, and reduce management expenses. In order to benefit from the collective knowledge of the city, a smart city links physical infrastructures with information technology infrastructures, social infrastructures, and business infrastructures [3]. IoT devices can be widely deployed in a city to make it smarter (especially through machine-to-machine and human-to-machine communications). IoT's sensing-actuation arm, Wireless Sensor Networks (WSNs), interact with urban infrastructure in a "digital skin," seamlessly. In order to establish a city's Common Operating Picture, the collected data is shared across several platforms and apps (COP) [4].

The Internet of Things idea uses various ubiquitous services to enable the development of smart cities throughout the world. It opens up new possibilities, such as the ability to remotely monitor and manage devices, analyze, and act on data from multiple real-time traffic data streams, thanks to the Internet of Things. To put it another way, Internet of Things (IoT) technologies are revolutionizing cities by upgrading infrastructure, developing more efficient municipal services, increasing transportation services by lowering traffic congestion, and enhancing public safety for their citizens The full potential of the Internet of Things (IoT) will only be realized if cities provide scalable and secure IoT solutions instead of just supplying a smart city component, which incorporates effective Internet of Things (IoT) technology.

Connected cities use telecommunications and information technology to improve the quality of life for their citizens. These cities are called smart. There are two main ways to make a city smarter: For the provision of a high-tech urban infrastructure capable of gathering and processing data via upcoming technologies, such as the smart grid and metering.

Reducing CO_2 emissions by enabling customers to interact with the environment through the Internet of Things (IoT). Reduced pollution levels help the environment and, as a result, the quality of life for citizens (e.g., improved health, safer, faster, and cheaper commute) [5].

Figure 9.1 shows the various components that make up a smart city. Data collection, transmission/reception, storage, and analysis are common components of smart city applications. The collection of data is specific to each application and has been a major motivator for the development of sensors across a wide range of fields. Second, data exchange is required, which comprises the transfer of data from data gathering units to the cloud for storage and processing Many smart city initiatives include city-wide Wi-Fi networks, 4G and 5G technologies, as well as different forms of local networks capable of sending data on a local or global scale, to achieve this goal. The third stage is cloud storage, which uses a variety of storage systems to store and organize data in advance of the fourth stage, which is data analysis. In data analysis, patterns and conclusions are extracted from gathered data to help make decisions. Fundamental decision making and aggregation, as well as other simple analysis techniques, could prove useful

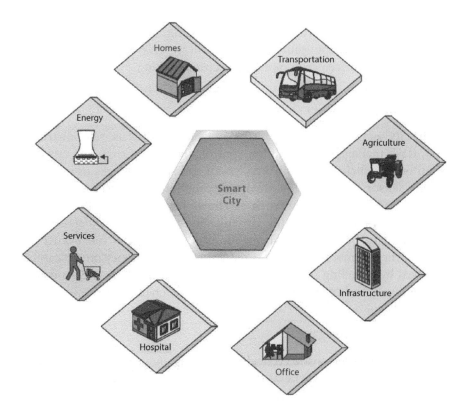

Figure 9.1 Components of smart city.

in some situations as well. When it comes to making more sophisticated decisions, the cloud's availability makes it possible to collect and evaluate heterogeneous data in real time using statistical methods, as well as using machine and deep learning algorithms [6].

9.3.1   Smart Agriculture

Food security is one of the UN's Sustainable Development Goals for 2030. Governments throughout the world have made it a top priority to guarantee that food production is sustainable and that finite resources like water are used wisely in light of the world's growing population and the worsening effects of climate change on the weather in agricultural regions. Smart agriculture is a burgeoning sector that uses sensors implanted in plants and fields to analyze a range of data to aid in decision making and prevent disease and pests. Sensors are utilized in plants as part of the smart agriculture paradigm to offer targeted measurements, enabling for the application of customized care mechanisms. Agriculture with a focus on precision. Precision agriculture is a key part of the attempt to boost sustainable food production in the future in order to assure future food security. AI in IoT for agriculture applications is important for crop monitoring/disease detection, data-driven crop care, and decision making.

9.3.2   Smart City Services

Smart city services include municipal obligations, such as water distribution, trash management, pollution control, and monitoring. The city can monitor the efficiency of its water system and spot any leaks using water quality sensors. Many of the previously mentioned smart city initiatives include waste management, such as chutes in Barcelona and bins connected smart and linked to the virtualization to notify the officials when they ought to be emptied even while utilizing artificial intelligence to identify the optimal route for cost reduction. Saving money on gas can be achieved by installing sensors that monitor pollution levels and send cars to the closest freely accessible parking place when necessary.

9.3.3   Smart Energy

The energy flow in most electrical networks is one-way, with the energy coming from the main generator, whether it is hydroelectric or fueled by fossil fuels. It is possible to regulate power generation by receiving feedback from the substations. However, because consumers do not provide feedback, the generation scheme employed with these systems relies on power production from these sources far outpacing demand so that there is always electricity available. In such systems, locating problems and taking corrective action might take a long time as well. Aside from receiving a supply from their primary utility, the modern consumer also produces their own electricity as renewable energy technologies become more inexpensive. Using information and communication technology (ICT), smart grids can make existing and new grids more observable, enable distributed energy generation for both consumers and utilities, and improve system self-healing. In recent years, smart grids have grown in popularity. One element of smart grids is the provision of real-time electricity data to utilities at multiple grid locations along the supply chain till the consumer. It is possible to better manage power generation by employing prediction models generated using consumption data gathered from smart grids, to integrate various energy sources and to self-heal the network in order to assure an continuous supply.

9.3.4   Smart Health

Information and communications technology is being used to improve healthcare quality and access under the umbrella term "Smart Health." As the world's population grows and healthcare expenditures rise, experts and healthcare professionals have turned their attention to this problem. We have overburdened healthcare systems that can no longer meet the rising demand from society. Intelligent healthcare strives to make healthcare more accessible by using telemedicine services and artificial intelligence (AI) to aid clinicians with diagnosis. Inertial sensors in smart phones and wellness trackers, which can record daily activity and detect abnormal movements while also collecting real-time health data (such as Electrocardiography, temperature levels, and body oxygen levels), have made it possible to use virtualized computation for universal healthcare actions. As a result, healthcare facilities will experience less financial and operational stress.

9.3.5   Smart Home

The smart home is a critical part of Smart Cities since it serves as the hub of daily life for city residents. In order to have a smart home, sensing units must be deployed around the house to gather data about the home and its residents. These sensors may include environmental detectors, surveillance cameras, and power sources usage monitors for the user's activities.

9.3.6   Smart Industry

Worldwide, businesses strive to improve efficiency and productivity while also reducing expenses. The Industry 4.0 model envisions a networked factory with all of its intermediary functionaries completely integrated and working together. This was made feasible thanks to the Internet of Things (IoT). When it comes to manufacturing and production, using the Internet of Things (IoT) has provided various benefits for industry, including faster and better innovation, modern manufacturing strategies (data, procedures), higher quality products and greater worker safety. As a result, smart industries confront a variety of IoT application issues when working with various devices and machines. Smart Industry IoT systems face a number of obstacles, including the need for adaptable setup, proximity, and speedy installation. Industry 4.0 services have been aided by AI and IoT, which have worked together to accelerate their development and adoption. Because of the sensors that are being used in the plant, data from these sources can be used to boost automation, execute business intelligence activities, and other things. Researchers have developed frameworks for incorporating artificial intelligence into the Internet of Things for Smart Industry. Machine health monitoring, problem detection, and production management are all key uses of artificial intelligence in the industrial sector.

9.3.7   Smart Infrastructure

In order to maintain a high standard of life, a city's infrastructure is critical. City governments must develop new bridges, roads, and buildings for their residents' benefit, as well as maintain existing ones. Structure health monitoring using accelerometers and smart materials is made possible by deploying sensors to measure the structural state of buildings and bridges for smart infrastructure. Predictive maintenance is possible because to the data collected by these sensors, which keeps the city's vital systems running smoothly.

9.3.8   Smart Transport

Congestion, pollution, and concerns with public transportation scheduling and cost reduction plague many cities. We are used to quick development and application of new technologies in the fields of information and communication. When it comes to vehicle-to-vehicle communication and infrastructure connectivity, such technologies have opened the door to the development of smart transportation systems. Vehicles can now communicate with each other, infrastructure can communicate with pedestrians and pedestrians can communicate with vehicles. In light of the fact that nearly all cars are equipped with GPS devices and that nearly everyone owns a cellphone, numerous systems make use of GPS data to monitor driver behavior and traffic patterns. Route mapping and public transit trip planning are currently made possible with the help of real-time data from apps like Waze and Google Maps. Sensor-equipped parking systems can also direct motorists to the next available place.

9.4   Smart Home Automation

A smart home system is anything that is connected to the internet and can be controlled remotely. A "connected home" is a term that refers to a house where all of the technology functions together as one unit. All of your home's electronic equipment, such as the thermostat, lights and audio speakers can be controlled via a single system that can be accessed via a mobile phone or a touch screen mobile device [7].

Electronic appliances and sensor status may be checked and changed by using a system for automating the home in the future, which is demonstrated. Lights, heat, ventilators, and climate control are examples of common appliances, as are other electronic devices that provide these functions. Aside from saving time, this product also offers energy-efficient solutions that reveal which of your home's devices uses the most electricity. This energy-saving and cost-effective solution can be used in hotels as well as in industrial and home settings. When you are away from home, you can stay in contact with your house thanks to the app's simple Notifications and a graphical user interface based on icons [8]. When compared to the cost of special IP-based (RJ-45) equipment, while the idea of monitoring already existing household gadgets is appealing, the actual implementation is much more expensive [9].

Figure 9.2 shows the functionality of this smart home system. On the admin side of the software, users may build their entire home's structure using a simple stretch interface. To begin, the user must first add floors to the house before selecting a floor to begin adding rooms to. As soon as the user has chosen a room, they will be able to add devices and customize their placement in accordance with a home's actual structure. Inside the house, you can also create your own floors, rooms, and gadgets if you so like. With their credentials, a user can examine the whole structure of their house on our server in JSON [10] form as it was configured in the admin application's sync database. That is because every 30 seconds, the server

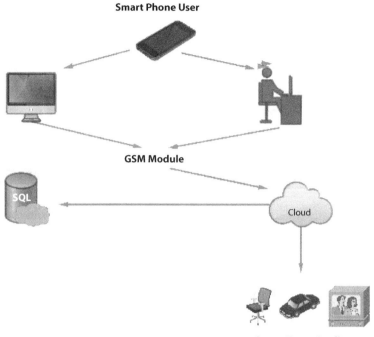

Figure 9.2 Working structure of home automation.

fetches all of the data and updates it with the VOLLEY service. The total number of devices installed in the home may be seen on the app's main screen, as well as floor information. There are three more tabs at the bottom of the app. The whole sensor status may be found on the second tab. The third section displays a summary of machine updates, together with the username who made the change and when it was made. The logout option can be found on the fourth tab. For credentials, shared preferences are maintained and the data is kept.

(i) To classify the status of smart home system equipment, the SVM classifier is also utilized. It categorizes home appliances as "ON" or "OFF" according on how they are utilized. We evaluated our current model to a slew of others to determine which was the best. One must keep in mind that IoT devices and their users are protected by blockchain technology, which can be used to turn appliances on and off. The usage of blockchain technology ensures secure data transfer between smart home automation devices, servers, and applications. Modeling of Variables in a Systematic Way. The smart home automation system under consideration is capable of intelligently deciding on the status of home equipment in its vicinity. We utilized a linear SVM classifier developed by Vapnik [11]. Svm is a supervised learning approach that may be applied to both regression and classification problems because of its versatility. We have made use of it to organise the system we are developing. Decision trees, neural networks, and K-nearest neighbours are among the data categorization methods. These are but a few instances among the countless possibilities. The following are the reasons we opted for SVM-based logical decision in the proposed approach: We needed binary classes, and SVM was the best choice. There are two types of gadgets in our system, and each one must be classed accordingly, i.e., "ON" or "OFF."

(ii)　The optimal hyperplane for classifying data points of one class from next class in cataloging issues must be found using a kernel function. Using SVM, the best hyperplane is found by separating both classes of data points.

(iii)　SVM classifier classification produced a hyperplane with a greatest margin between target class data points (iii). Other classifiers' hyperplanes, on the other hand, fall short of the ideal [12].

Temperature, smoke, and light sensors are included in the input dataset of our problem. Grade classes, i.e., "ON" or "OFF," are the problem's output. By setting $y = 0$, we indicate that the device is "OFF." By setting $y = 1$, we indicate "on."

In order to provide a reliable and secure identification and authentication of IoT devices, the smart home automation system is a major objective. We have employed blockchain technology to make sure these goals are met. In 2008, Satoshi Nakamoto released the first version of the Blockchain software [13]. Security, anonymity, and decentralization are the three fundamental benefits of blockchain technology [14]. More security and less reliance on the central server are two benefits of using these features in IoT devices. The use of data encryption and timestamp in blockchain technology ensures tempered resistance data structure as well. The blockchain module is implemented in Java using the recommended approach by specifying the block's contents in a hash that serves as a unique identifier for each block. A block encryption is produced for each block, and a SHA-256 hash is then derived from it. A block is formed when a certain threshold is reached and a request for connectivity is granted. This is done through the management of the blockchain. To verify the whole blockchain, a block from the chain is looped around to verify the hash value of the current block against the value of the prior block.

Figure 9.3 shows the flowchart used to authenticate a connectivity request whenever something is generated by the user. To understand the working of the project, Figure 9.2 explains the complete working structure and integration of different devices with each other. The arrows show the flow of projects starting from user smartphones to changing the state of electronic devices. There are two network modes in which the user can interact with our Raspberry. If the user is sitting inside the home (home local network), then the user will be able to use all of the IoT services at the local network without connecting to the internet cloud. This will also result in faster communication of devices with the user as everything is happening locally.

The second network mode is used if the user is residing outside the home anywhere in the world. Then, the user first connects to the internet. The processed request is sent to the Microsoft Azure Cloud. Based on credentials provided by the user matched with the Azure cloud database, the user request is sent to the respective Raspberry for processing. The account of each user is maintained individually at the Microsoft Azure Cloud Database. Services of each user are handled based on credentials from which the request is being generated. APIs are called from the cloud if the user is outside the home network. Then, the same APIs are also residing inside the Raspberry Pi server if the user is in the home network.

Data sharing between application and server database is performed in the form of JSON. APIs are secured using multiple hashing techniques. /e change in the state of any device is performed using Raspberry Pi GPIO pins. Raspberry Pi receives the request from the server. According to the request from the user, Raspberry Pi responds to devices. /e database of each request generated by the individual user is maintained at cloud servers. /e user can check the complete history of processed requests on his smartphone by setting the duration. Sensors installed inside the home update its state continuously after 30 sec and respond to the change to the Raspberry Pi server. In response, the Raspberry Pi server syncs all data to the cloud database, and the values of the mobile application are updated.

This system comprises two modes, the admin side in which the user is able to design the complete prototype of the home and the user side in which the user able to control each device of the home with an easy GUI-based interface. This system also has decision-making ability about

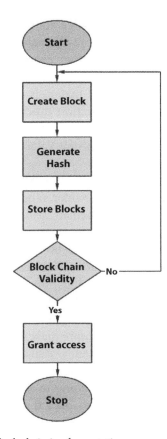

Figure 9.3 Flowchart of block chain implementation.

the status of each device of the home. A machine learning algorithm SVM with a linear kernel is applied for decision making about the status of home appliances either ON or OFF. The presented approach also ensures secure identification and authentication of IoT devices using blockchain technology, while the existing systems which have not copped with intelligent decision making and analytical abilities are the need of time.

9.5   Automation in Manufacturing

IoT and smart manufacturing are at the heart of Industry 4.0's core principles. Effectively interact along the entire product life cycle may interact and supervise one another on their own, with far less involvement from operators, when machines are linked together. Instead of depending on inconsistent monitoring by maintenance staff, Bosch, GE, and Johnson Controls, the Internet of Things (IoT) will enable robots to automatically detect failure and begin repairs. Another example of Internet of Things (IoT) use is self-organized logistics, which reacts to unforeseen manufacturing changes like shortages of materials or bottlenecks. Dynamic, efficient, and fully automated production processes will be provided by manufacturers thanks to the application of technology. By linking various devices, the Internet of Things creates new value:

- People. While high-value processes presently benefit from embedded devices, a much wider range of manufacturing processes will be able to use them in the future. Manufacturing companies can better serve their customers by connecting them with relevant information on the user's device at the proper time, and throughout organizational units to include vendors, servicing partners, and transmission chains, with the use of Internet of Things (IoT) in manufacturing technologies. For a fraction of the cost of earlier specialized systems, new mobile-ready software makes it possible to access embodied energy, line efficiency, visual analytic tools, and warnings from virtually any location.

- Process. Manufacturers are going to look for specific visibility and supply in order to plan their operational challenges in the early stages of IoT deployments. Third-party managed solutions may be deployed by manufacturers to get started. Manufacturing will be able to respond more quickly to the market as IoT grows more widespread by linking devices into operational and corporate software processes. New levels of automation will be possible because to M2M communications. When it is too moist to paint a car, for example, General Motors consists of sensors data to figure out that. Vehicles are sent to another section of the manufacturing process if conditions are not favourable, which reduces repainting and maximises plant uptime. The corporation saved millions of dollars as a result of just this one modification.

- Data. Newly linked gadgets will generate new sorts of data as a result of mobility and the Internet of Things (IoT). Sensors, actuators, video cameras, and RFID readers will all be part of the Internet of Things (IoT). Data from IoT-enabled devices will be collected and analysed using on-premises or cloud-based big data processing and analytics. These innovations will help people and machines make better judgments by converting data into context [15].

9.5.1   IoT Manufacturing Use Cases

The Internet of Things has an impact on every industry, but it has the potential to alter industries like manufacturing, utilities, and aviation. Large and little volumes of data will be transmitted by an increasing number of machines and gadgets. Analytics will be used by smart manufacturing businesses to make better decisions and run their operations more efficiently. IoT, big data, and IP networks will allow firms to extend the life of their assets while also boosting efficiency and using less energy. mart manufacturing systems will incorporate Material Requirements Planning, Manufacturing Resource Planning and Manufacturing Execution Systems. IoT has numerous potential business applications in the manufacturing industry. Several instances of how the Internet of Things (IoT) can improve processes include:

- Accessibility to the factory. The Internet of Things (IoT) and IP networks will link what happens on the factory floor to enterprise-based systems and decision makers. The Internet of Things (IoT) will give decision-makers access to information about the production line and increase factory efficiency. Using IoT and visibility technologies, a plant manager strolling the production floor, for example, may access the efficiency of any machine, see production from any place, and cut down on the time it takes to make a decision and implement change. Tablets may now display performance data and status updates previously available only on PCs with GE mobile-enabled SCADA software.
- Facilities managers and production staff will no longer be confined to a control room and will be able to work more efficiently thanks to simple access to real-time information. The advantages of visibility spread yonder the company to a extensive assortment of service providers, consumables, and capital goods suppliers and third parties. Because of the enhanced visibility and remote monitoring afforded by IoT systems, third-party suppliers will be able to get much more involved in the direct operations and maintenance of industrial facilities. If equipment output and maintenance status

can be monitored adequately, There is a possibility that major equipment suppliers will now be in a position to offer marketing strategies that involve remuneration based productivity rather than sale of productive assets. Maintenance, Repair and Operations (MRO) suppliers will leverage Internet of Things (IoT) to monitor scattered stocks, fluid tank levels, worn part conditions, and production rates. Because of this, manufacturers and suppliers will form whole new kinds of commercial partnerships that are intricately interwoven.

- Automation. Networks of plants have been cut off from one another and from nearby and far-off commercial networks. There are many ways to connect things in a plant using IoT and IP networks nowadays. Using the information gathered from connecting machines and systems in the plant, manufacturers can use workflow automation to maintain and improve production systems without involving humans. Harley-York, Davidson's Pennsylvania motorcycle manufacturing, for example, makes extensive use of the IoT. As an example, in the painting booth, the company installed software that records the speed of fans. Software can adjust the gear if a parameter (such as blade speed, temp, or moisture) deviates from the set limits.

- Energy management. Energy is usually the second-largest operating expenditure in many enterprises. To make the most of energy use across a wide range of different industrial operations and locations in real time, many companies lack the essential measuring instruments, modelling techniques or management and performance tools. Automation of environmental controls like HVAC and power, as well as the Internet of Things, can save manufacturers money in multiple ways. With interconnected energy systems, peak - load charges can be eliminated, and business model activities can be carried out. The integration of meteorological data and forecast analysis into certain IoT-enabled HVAC systems can help firms better understand their costs and better plan their energy usage.

- Proactive maintenance. Preventative and condition-based monitoring is widely acknowledged by manufacturers, but many are still in the process of putting it into practice. Data collection and equipment health monitoring are now more cost-effective and easy thanks to the availability of lower-cost sensors, wireless connectivity, and tools for large-scale data processing. Actively monitoring equipment out of range with sensors and thwart problems is an option for manufacturers who have equipment that is required to work within a specified temperature range. The measurement of vibrations to find out if a machine is operating outside of specification is another example. Businesses, particularly industrial ones, incur financial losses as a result of malfunctioning equipment. Using new sensor information and the Internet of Things (IoT), a manufacturing facility can improve overall equipment effectiveness (OEE), save money by minimizing equipment failure, and perform planned maintenance.

- Connected Supply Chain. Analytics, IoT and IP networks will aid producers in understanding real-time information with suppliers. The concept of just-in-time manufacturing is not new. By connecting the production line and balance of plant equipment to suppliers, all parties will have a greater understanding of interdependencies, product flows, and production cycle times. Devices with IoT capabilities can be used for a variety of purposes, including location tracking, remote inventory health monitoring, and tracking on parts and items as they travel through the supply chain. The information gathered by IoT systems and entered into ERP systems can be used to give up-to-date billing information. With real-time information access, it will allow firms find potential issues, reduce their inventory holding costs, and maybe reduce their capital reserves as well.

9.5.2   Foundation for IoT in Manufacturing

Smart manufacturing is built on a foundation of at least four different technological factors. Some features are as follows:

- Network: According to Cisco data, only 4% of production floor equipment are truly networked. Proprietary networks have been employed by many manufacturers in the past. An IP-centric network is necessary for a smart manufacturing environment because it allows all devices in a plant to communicate with operational systems as well as enterprise business systems. IP networks standardise connectivity and collaboration across the supply chain, increasing visibility throughout the entire chain. RF problems in plants, extreme climatic conditions, and alarm transmission reliability and real-time data stream processing necessitate robust networks for manufacturers.

- Security: The most frequently mentioned impediment to the implementation of smart industries was a lack of IT protection. Managers of operations must verify that measures such as hardware confidentiality, structural building security, and network monitoring for data in transit are included into the solution. In addition, secure global access to the network must be possible via the network. To endure tough climatic conditions not found in ordinary networks, security and connectivity solutions must be configured to withstand heat and moisture as well. Structures for identity and authentication will also have to be changed to include "things" as well as people.

- Software systems. This new IoT data is distinct from what we have been using to run our systems in the past. Data from a network of sensors is needed to complete the task. Models and software must be utilized to translate information from physical world and turn it into actionable insights that can be put to use by humans and robots alike.

- Big data and analytics. Despite the fact that manufacturers have been creating large amounts of data for some time, they have only recently been able to store, analyze, and make use of all of this data. There are new methods for analyzing large amounts of streaming big data in real time, allowing for huge advances in issue solving and cost avoidance. Forecasting, preventative maintenance, and automation will all be built on top of big data and analytics.

As a result of the Internet of Things, the way items are developed, manufactured, distributed, and sold will radically alter. Leading manufacturers create their products and services to be flexible and evergreen. It necessitates a radical change in the way things are designed, as well as the factories and systems that support them. Manufacturers may become more efficient, increase worker safety, and create new business models by utilizing IoT, IP networks, and analytics technologies. The Internet of Things (IoT) will aid firms in increasing resource efficiency, enhancing safety, and maximizing the return on their investments. Manufacturers who learn to work with this new dynamic will find new ways to boost their income and save money.

9.6   Healthcare Automation

In order to live in a society that is healthy, social, political, and energy-efficient, we must be aware of and use latest innovations in exceptional standard health systems driven by the Iot devices (IoT). A system or eco-system can be transformed for the benefit of people and the earth if bear-able growth, energy effectiveness, and community health are taken into consideration as interrelated factors. Sustainable development objectives should be realized thanks to the use of smart devices that incorporate sensors.

The Internet of Things is a significant factor in the evolution of health-care from traditional to smart healthcare. i) In order to provide smart healthcare, the following things must be done: the right action should be given to the right patient at the right time. ii) It aids medical profession-als in accurately diagnosing and treating patients. All parties are able to share information and communicate effectively as a result of this. Provide patients with timely information so they can be actively involved in their own care. iii) Make data freely available and easily accessible. iv) Due to its cost-effective models, healthcare may now reach people in far-flung areas. v) Reduces waste and operating costs, which increases efficiency.

Hospitals, pharmaceuticals, medical devices, and medical supplies make up the majority of the healthcare market. Telemedicine is a form of remote monitoring of patients with the use of medical devices and equipment. By 2022, the healthcare industry is anticipated to be worth $372 billion. Sedentary lifestyle diseases account for roughly half of the money spent on within beds in major cities. Hypercholesterolemia and increased blood pressure are both linked to being overweight or obese, and alcohol intake is a major risk factor for these conditions. Telemedicine's IoT is advanc-ing quickly thanks to advancements in Internet and telecommunications-related technologies. Rural and urban healthcare can be linked together using telemedicine. Since it is inexpensive to consult with and has access to remote diagnostics, it is ideal for people who live in remote locations. Home healthcare is another use for the Internet of Things. As IT and medical device integration progressed, as a result, patients could receive significant, low-cost healthcare at home.. Patrons can expect to save somewhere between 20 and 50 percent on their purchase price. Mobile computing is one of many technologies that fall under the awning of the IoT. Customers can now get healthcare services at lower costs because to improvements in mobile technologies such as 4G and 5G in health-care [16]. Since IoT has had such a substantial impact on the healthcare industry, this paper's major contribution is to examine several elements of IoT in healthcare, including:

    i.   Smart universal healthcare empowering technologies.
   ii.   Internet of Things in Healthcare
  iii.   IoT-based healthcare apps with a simple and minimal architecture
  iv.   Obstacles and current solutions

9.6.1 IoT in Healthcare Applications

Figure 9.4 depicts some of the many IoT-based healthcare applications that have been discovered. Apps that have the same scope and focus are classified together under the following categories: • Telehealth; • Chronic illnesses diagnosis and treatment; • Home healthcare for the seniors; • Real-time surveillance with alarm production. After that, we will go over the many sorts of IoT-related healthcare systems and their roles in each of the sections.

Real-Time Monitoring and Applications

Real-time healthcare services rely heavily on the constant monitoring of various health vitals like body heat, heartbeat, and oxygen content. This monitoring is critical. In the Internet of Things, sensors can be placed on

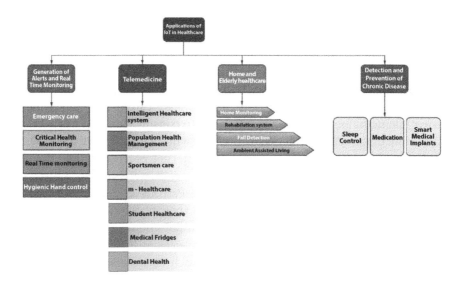

Figure 9.4 IoT applications in healthcare.

people's bodies, and various metrics can be measured as a result of this [17]. This information can be used to provide patients with emergency warnings with prescription recommendations based on an analysis of the data. ECG, temperature, foot pressure, and heart rate may all be continuously monitored with an e-health system created by Shanin *et al.* [18]. Flexibility and low power consumption are two of the advantages of this technology. To provide emergency healthcare, the GPS location of patients is tracked. RFID tags are utilized to identify specific patients. The Microcontroller Unit (MCU) was an Arduino Uno, and the middleware was thingspeak. Swaroop *et al.* [19] created an amazing health monitoring system using a DS18B20 digital temperature sensor and a Combined blood pressure/ heart beat sensor. Heat and blood pressure were measured and communicated using a variety of transmission channels, including Bluetooth Low Energy, GSM, and WiFi. As a result of the various steps that data must go through, BLE was shown to have a lower bandwidth than WiFi. Using the Apache Spark and Hadoop ecosystems, Rathore *et al.* [20] presented a real-time emergency response system that processes enormous volumes of sensor-generated data in order to reduce latency.

Telemedicine
Telemedicine is the practise of giving medical care to patients over long distances via the Internet and other forms of electronic communication. This lowers the operational costs for medical workers while also improving the health efficiency of the patient. By combining neural networks and fuzzy systems, Zouka and Hosni built a smart healthcare monitoring system that examines sensor data such as blood pressure and pulse oximeter readings like an electrocardiogram (ECG). The Fuzzybased Inference System and Azure IoT Hub process sensor data given to the Azure IoT IoT Hub (FBIS). Doctors can now provide emergency healthcare to their patients using an M2M patient monitoring gadget and a remote health app. Rohokale *et al.* [21] developed a rural healthcare monitoring system using vital signs such as blood pressure, sugar levels, and abnormal cellular growth. The patients wear an RFID tag as a means of tracking their whereabouts. Remote healthcare facilities receive notifications if any of the patient's health indicators are abnormal, and the professionals there will respond accordingly.

Chronic Disease Detection and Prevention
Chronic diseases such as cancer, diabetes, asthma, and obesity affect a large number of individuals around the world. Depression is a typical consequence of chronic illness caused by certain illnesses. Patients with

autism may benefit from a health monitoring system created by Sundhara Kumar *et al*. Neurosensors are used to collect data, and the system uses it to read EEG waveforms and notify caregivers if there are any odd findings. Doctors who watch their patients from a distance receive notifications in the event of an emergency through email. When it comes to enabling IoT-based healthcare solutions for cancer patients, Onananya and Elhakankiri [22] suggested a variety of designs and frameworks. Large-scale sensors are supported by wireless sensor networks (WSNs), a communication system that uses big data technologies for analysis. It has been brought out how IoT healthcare systems face operational and security challenges. Using IoT and cloud computing, Sood and Mahajan [23] designed a health system to identify the chikungunya outbreak. Using sensor data and fuzzy c-means, fog nodes can detect illnesses in users. Immediately, users and experts receive notifications. Information about a patient's general health and medical condition is stored in a highly secure cloud. To discover outbreaks in a certain location, social media platform analysis is also performed. Chikungunya epidemic notifications are sent to the appropriate public and private healthcare organizations, which aids in improving service and halting the spread of the disease.

Home and Elderly Healthcare

IoT and associated technologies can be installed in the homes of older people who are unable to move quickly and require regular or emergency healthcare assistance. Abdelgawad *et al*. [24] have proposed health monitoring systems for active and supported living that use a variety of sensors to obtain data and transfer it to the cloud for analysis. BLE-enabled indoor location modules and sensor interface circuits were used in conjunction with these sensors to show their system in action. In addition to WiFi transmitters and cloud servers, the Raspberry Pi 2 served as their microcontroller. To determine the prototype's effectiveness, experts will look at how it performs in real-world situations. Yang *et al*. propose a personal healthcare system for wheelchair users who are alone at home. To provide high-quality healthcare to wheelchair users, this system makes use of the Internet of Things and Wireless Body Sensor Networks (WBSNs). Sensors measuring pulse rate, Electrocardiogram, stress, optical remote sensing and actuators are put on and around a patient to keep tabs on his or her health in real time. WBAN collected data from the patient and the environment using a mobile phone as a gateway and Zigbee and Bluetooth connection protocols. Environment sensors in the wheelchair may track its location as it moves from one location to another, addressing another aspect of patient mobility. Patients' vital signs can now be tracked via the Internet of Things, thanks to new research from Yang *et al.* [25]. Multiple ECG sensors were attached to the patients, allowing the researchers to keep tabs on their vital data. The sensors gather information and send it wirelessly to the cloud. The Internet of Things (IoT) cloud is made up of high-performance computers that can process and analyze data to produce actionable results. An online Graphical User Interface (GUI) was made available for displaying cloud-based data.

9.6.2   Architecture for IoT-Healthcare Applications

Providing healthcare to people who have been involved in life-threatening accidents or illnesses necessitates the prompt involvement of medical personnel. This type of event necessitates a rapid and decisive response with minimal latency. The delay for data transmission to the cloud, processing in the cloud, and receiving a response is high in a generic cloud system, and this is unacceptable. Fog computing, which delivers processing and storage resources to the network's edge, or closer to the sensors, can help us solve or limit the latency concerns. The majority of healthcare solutions currently in use make use of cloud computing to aid in decision-making. Fog computing for time-critical healthcare applications has been proposed in more and more solutions in recent years.

Each of the levels of the fog-based architecture has its own function, and together they form the fog-based architecture. The sink layer adds latency to data, and the fog layer adds complexity. There are latency difficulties with cloud-based solutions. This method addresses them. A discover the underlying, a foggy surface, and a cloudy surface are shown in Figure 9.5, respectively. This architectural style may accommodate both time-critical and non-deadline-driven applications such as rapid response and mobility emergency treatment. The sensing layer has various sensors for keeping track of patients' and the elderly's vital signs. Mobile devices, Arduino boards, Raspberry Pis, and other hardware platforms can access the data collected by the sensors through Bluetooth or USB. The new information is then promoted to the mist layer by these application devices. Depending on the application, data can be filtered or processed on one or more transitory storage servers in the fog layer. Local gateways can also enforce security and privacy standards. A border gateway receives the data once it has been transformed and sends it to the cloud. Malicious traffic can be filtered using a firewall at either the local or border gateway. We take it for granted that data exchanges within the network take place over encrypted channels. Depending on the use case, data can be stored permanently in the cloud or reviewed by medical professionals for decision making [18].

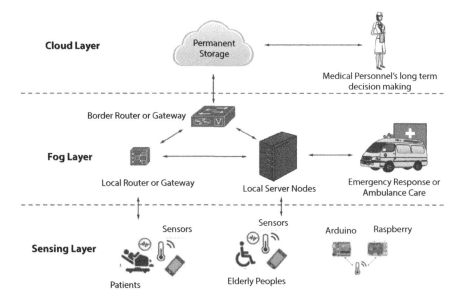

Figure 9.5 Fog-based architecture for healthcare applications.

9.6.3   Challenges and Solutions

There is a great agreement of study being done to address a variety of issues connected to the use of IoT and related healthcare technologies. Nevertheless, there are numerous unresolved issues [26]. We will look at some of the most pressing issues facing the industry right now in this section as well as the solutions that have already been proposed.

Fault tolerance
Whether data is sent to the cloud or processing layer, it affects the Remote health monitoring system's resilience. Their operation is critical to the efficient and effective delivery of healthcare in times of crisis. Only an IoT design with standby unit for communication breakdown was presented in the current literature as a solution for fault tolerance. This solution's gateway diagnoses sensors on a regular basis by looking at data trends and taking relevant action.

Latency
The healthcare application's efficiency is heavily reliant on the transport and communication layers' network latency. The impact of delay varies by healthcare application type as well. Rescue services and interactive media monitoring devices like live surgery would be significantly impacted by latency. A fog-driven IoT healthcare design reduces communication

latency by conducting pre-processing steps close to the sensors at the portal rather than sending data to remote, which adds additional complexity.

Energy efficiency
The perception layer's sensors are all powered by batteries in some way. When the battery runs out, sensor nodes must be turned off to keep the system running in an emergency healthcare scenario. An independent WBAN deployment for healthcare has been suggested that uses sun's radiation rather than battery-powered nodes. The panels were exploited to their full potential, and the system's ability to operate 24 hours a day was demonstrated through testing.

Interoperability
The Internet of Things (IoT) connects disparate fields of study by bridging disparate standards. This creates an interoperability challenge. The healthcare industry has a simpler time dealing with this issue because of the strict laws and standards already in place. IEEE 802.15.4 is an open standard that was established on top of 6LoWPAN to overcome interoperability difficulties among communication nodes. A real-time health monitoring system based on this standard has been proposed. Middleware abstraction of complex details has been proposed as a solution to interoperability concerns. There was a proposal for an IoT semantic interoperability paradigm to deal with the difficulties between heterogeneous IoT devices. RDF and SPARQL are used in this architecture for associating objects and querying them to get information.

Availability
Physicians and clinicians can access a patient's data in IoT healthcare apps such as actual surveillance and critical care services around the clock. A patient's mortality could be caused by a lack of crucial data if nodes or networks are unavailable at the observation or definite path. A project named VICINITY, which received funding from the EU, is able to adjust the communication load based on the situation. More research is required to protect the processing infrastructure, which includes cloud computing, big data, and other cutting-edge technologies. Many dangers and vulnerabilities exist in cloud computing that need to be addressed if an IoT healthcare system is to be highly available.

9.7   Industrial Automation

Interconnected sensors, controllers, and other industrial equipment are often stated to as the IoT, which is a general term. Because of this connectedness, users may access and monitor their systems from anywhere, and they can also collect, exchange, and analyze data from a variety of various sources. The Industrial Internet of Things (IoT) has huge productivity, cost-saving, and efficiency-improving potential. Low-cost and quick installation distinguish IoT solutions.

9.7.1   IoT in Industrial Automation

In the field of industrial automation, commercial computers, Programmable controllers, and other computation control systems or robots are used to automate industrial processes and machines and eliminate the need for human intervention. Increasingly, the internet of things in industrial settings plays a critical role in industrial automation.

Through IoT, industrial automation may generate more efficient, cost-effective, and customer-specific solutions. It can have a significant impact on the efficiency and uptime of industrial equipment (such as PLCs, robotics, actuators, and sensors) by connecting them to the cloud and sharing real-time data [26].

9.7.2   The Essentials of an Industrial IoT Solution

While the Internet of Things (IoT) aims to connect everything digitally, the Internet of Things (IoT) concentrates on a specific region. Designed for service industries and organizations, IoT applications need to be more reliable and precise than standard IoT solutions in terms of robustness, flexibility, security, interoperability, reliability, and serviceability. Figure 9.6 depicts the fundamental IoT requirements.

Standard IoT solutions consist of the following:

- Industrial "things"—Internet-enabled devices such as PLCs, IPCs, Human Machine Interfaces (HMI), robots, vision cameras and sensors are typical IoT solutions. IoT solutions include IoT devices.

Figure 9.6 The essentials of Industrial IoT.

- Wi-Fi, 4G/cellular and ethernet connections all fall under the umbrella of connectivity.
- In the IoT, the value is in the data collected, stored, and processed by edge devices.
- Data hosting and remote services are made possible by a centrally located and secure cloud platform for IoT.
- Monitoring and data analysis with an analytics dashboard.
- To convey alerts or triggers to other systems, the data can be analysed by humans or intelligent operations.

9.7.3   Practical Industrial IoT Examples for Daily Use

Examples of IoT applications utilized in various industrial computerization circumstances are provided in the following real-world IoT examples. Investigate how the Internet of Things (IoT) is being used in your business today [27].

Remotely solve PLC/robot problems if a custom-built machine is down
Accidentally pressing the emergency button without realizing it happens all the time in factories. Because there is no flaw, engineers spend a lot of time scratching their heads to figure out what is wrong. The clock continues to tick and significant time and resources are being squandered as a result of the outage. If the HMI is not able to tell you what is wrong, you should contact the company that built the machine.

If necessary, the machine maker can reset the machine from their office using industrial remote access to see what has been going on with the PLC or robot. A quick journey to the factory eliminates the need for a lengthy service visit by locating the issue remotely.

Prevent the label printer from running out of paper
Running out of labels on a machine can be disastrous in the logistics or packaging industries. To avoid this, service personnel or operators must be alerted well in advance of when something like this is likely to occur.

As soon as the device's information counter reaches zero, an alarm is sent out to warn the user. This notification gives the operator the opportunity to respond quickly. If they get a push notification or email alert on their smartphones or a buzz on their watches, the responsible folks will get the message on time. Such alert signals are life-saving in various fields.

Publish New Functionalities on the HMI Screen for Customers Abroad
Depending on how your customer plans to utilise a machine after it is
delivered, he may want additional features to make his job even easier. An
on/off toggle or a percentage tracker for the motor can be simply added to
your program's control panel by your programmer. Upgrades and comprehensive testing of the HMI software are required after that before the new
features can be made available.

A secure network connection can be used to update the HMI software
remotely. Customers will be delighted once more when you just upgrade
your computer's software over the internet. A web-based VNC connection
allows you and the customer to inspect and test the IoT platform's HMI
capabilities on a mobile device Virtual Network Connection.

Predict Machine Maintenance and Analyze Upfront Which Part Needs To
Be Replaced
Maintenance is required for industrial devices and energy items like solar
panels. You can estimate when maintenance is required when you know
the deterioration over a certain number of processing hours or revolutions.
These scenarios call for the implementation of proactive maintenance and
the generation of accurate data to help drivers make better performance
decisions.

Using commercial technologies like as OPC-UA, Modbus, Siemens S7,
Ethernet IP, and so on, begin by recording data from your PLC's variables
(counters) to the cloud (see list below). Once you have done that, you can
either construct an IoT dashboard with data visualization or set up an email
alert to notify you whenever the clock strikes a certain servicing level.

If you know about the problems with your computer before you leave,
on-site maintenance visits will be more successful. If you use wireless
access and the web server's virtual diagnostic testing tool to investigate any
problems, you will have a better chance of having the right spare components on hand throughout the life of your implementation.

Analyze and optimize industrial robot actions
Repetitive tasks are simplified thanks to industrial robots like the UR+.
Robot programme actions can be remotely changed for changeovers or for
troubleshooting by using remote access and IoT features. Video analysis
can also assist a robot's actions be improved. Improvements are made simpler by having access to IP camera footage or live streaming. Make use
of the Hololens or other AR/VR technologies like a VPN connection to
quickly and simply gain network access to your robot controller.

Live monitoring of full garbage containers in smart cities
There will be no more pointless driving around town looking for empty containers to inspect. Only empty trash cans that sound an alert to let you know they need it. Use your sensors to their full potential and store data on the cloud. Then, when the container reaches a certain level, send a notification to the garbage collector using a monitoring dashboard.

Manage Data From Multiple Buildings for Central Monitoring in your BMS System
To monitor and operate numerous sites' energy usage, heating, lighting, fire protection systems from a central location, the IoT is utilized in building automation. Data from remote installations must be accessed to gain a clear picture of the HVAC system's state (Heating, Ventilation, and Air Conditioning).

By utilizing BACnet or Modbus protocols, edge connectivity aids in the communication of real-time machine data to a cloud-based central application. It is possible to exploit the growth of open cloud platforms for custom applications. Most of these systems have an API that allows you to gather data at predetermined intervals and send it to your BMS for centralized monitoring.

Automated and Remote Equipment Management and Monitoring
Many IoT applications are related to equipment automation, making it possible to operate and monitor all business operations from one place. Due to digital machinery and software's capacity to remotely operate equipment, it indicates that several plants located in different geographic regions can be controlled simultaneously. In addition to being able to view real-time progress in their production, organizations also have unprecedented access to historical data about their processes. This data is being collected and used to help support process improvement and create a climate where making informed decisions is a top concern.

Predictive Maintenance
Detecting the need to maintain a machine before an emergency arises and production must be immediately halted is known as predictive maintenance. Thus, a data collecting, analysis, and management system is one of the grounds for its implementation.

There are several Industrial IoT apps that work with sensors that can deliver alerts when certain danger indicators occur. This system is one of the most effective. Data is sent from robot or machine sensors to platforms, which evaluate it in real time and apply sophisticated algorithms that can provide alerts when temperatures or vibrations rise above usual limits.

Faster implementation of improvements
As a result of IoT, industrial business models (process, quality, or manufacturing engineers) can access and evaluate data more quickly and automatically, as well as remotely make the necessary adjustments to operations. This also speeds up the implementation of modifications and enhancements in Operational Intelligence and Business Intelligence, developments that are already providing competitive advantages to a wide range of industrial firms.

Pinpoint Inventories
Incorporating Industrial IoT systems makes it possible to check inventory levels automatically, verify compliance with plans, and send out alerts when deviations occur. Industrial IoT applications like these are critical for a smooth and efficient production process.

Quality Control
Monitoring the quality of manufactured products at each level, from raw materials used in the process to the way they are transported (through smart tracking systems) to the reactions of the end consumer once the product is received, is another key IoT application entry..

This data is crucial for evaluating the company's efficiency and making the appropriate adjustments if failures are discovered. This will help optimise operations and catch production-chain problems early. It has also been shown that protecting businesses like pharmaceutics and food is critical.

Supply Chain Optimization
The capacity to obtain real-time in-transit information about the state of a company's supply chain is one of the Industrial IoT applications targeted at increasing efficiency. In this way, various hidden chances for development can be discovered, and faults that inhibit processes and make them inefficient or unprofitable can be pinpointed and addressed.

Plant Safety Improvement
IoT-enabled machines can generate real-time data on the plant's status. Equipment deterioration, plant air quality, and the incidence of illnesses inside a corporation can all be monitored to help avoid potentially dangerous situations for employees.

This increases not only the facility's safety but also its productivity and motivation of its employees. Poor management of company safety also has economic and reputational implications, which are underestimated.

9.8   Automation in Air Pollution Monitoring

As a result, air pollution has emerged as a major issue. Due to the release of harmful chemicals into the atmosphere, such as CO_2, SO_2, NO_2, and CO, our planet is becoming increasingly polluted. These hazardous chemicals are dissolved in the atmosphere and are therefore unpredictable. As a result, a device is needed to monitor the air quality. Internet-based devices, such as IoT, can be used to keep tabs on pollution levels in the air. Data can be gathered via Internet of things (IoT) sensors and analysed depending on that data for predictions, such as if the air quality is good or bad. Local air quality can consequently be monitored using IoT-based devices such as sensors based on Arduino/Raspberry Pi. In order to provide data on contamination facts on the atmosphere, it is vital to understand environmental factors and enable for smooth installation into any other network that enables sensors to gather data on green technology eco system criteria [28].

Air is a necessary part of the human environment. Nitrogen, Oxygen, Carbon Monoxide, and even traces of other rare elements can be found in the earth's atmosphere. Humans require a contaminant-free environment for breathing. This is critical to the survival and health of human beings. Any alteration to the air's natural composition has the potential to be extremely harmful to all life forms that inhabit the planet. A pollutant in the air is present when there are large amounts of one or more gases in the atmosphere that are harmful to humans, animals, or plants. ppm or ug/m3 are units for measuring air pollution in parts per million or volume per million cubic feet. Direct releases of primary pollutants into the atmosphere occur. When a primary pollutant combines with other chemicals in the atmosphere, secondary pollutants are created. The state of the air has an impact on people's health. Air pollution can cause a variety of health problems, including wheezing, coughing, and worsening of asthma and emphysema. Visibility might be harmed as a result of breathing in polluted air. Worldwide, air pollution is responsible for 7 million premature deaths per year, or one in every eight people. Indoor/outdoor pollution and secondhand smoke are associated to about 570,000 paediatric deaths every year. Children who grow up in polluted environments are more likely to develop chronic respiratory diseases like asthma. There have been several

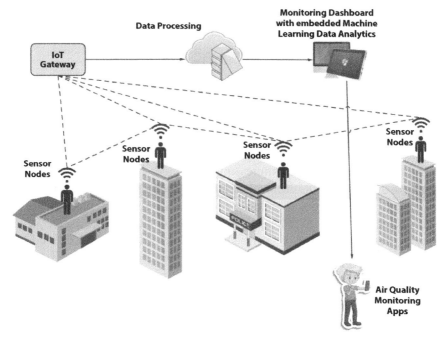

Figure 9.7 Air pollution monitoring.

models developed by scientists around the world to track a wide range of pollution compounds, including Sulfur Dioxide and Nitrogen Oxides. The goal of this project is to develop and implement a better air pollution monitoring system than what is currently available. It explains how a gas sensor, Arduino microcontroller, and WiFi module can be used to monitor pollution levels in the air. Main goal is to design an intelligent monitoring system for air pollution to monitor, analyze and log data on air quality to a remote server, and to maintain this information on the internet (see Figure 9.7) [29].

9.8.1   Methodology

It was decided to divide the system overview method into five tiers. The first layer was made up of environmental measurements and the results. This was followed by an examination of the sensors' properties and features as the second layer. Threshold valve fixing and regularity of susceptibility were all part of the third layer. This layer also included selection, detecting, and measuring capabilities. The sensor data collection was the job of the fourth layer. The ambient intelligence environment was the fifth layer. When the microcontroller operated the sensor, it collected data and sent it to the Wi-Fi module, which forwarded it over the internet for analysis. Measured parameters might be viewed by users on their mobile devices.

9.8.2   Working Principle

A message was sent to the LCD from the Arduino's library after it had been loaded up. Using the MQ135 sensor, researchers gathered information on air quality. The analogue output voltage was proportionate to the polluting gas concentration in ppm thanks to the calibrated sensor (ppm). The data is shown on the LCD screen before being relayed to the Wi-Fi module. Using the internet, the Wi-Fi module sends a copy of the measured data from the valve to the server. In order to send data to an application on a remote server named "Thing talk," the Wi-Fi module has been set up to transfer measurements. The online application allows anyone with an internet-capable device to access the measured data from anywhere in the world. The sensor data was turned into a string and delivered to the remote server, where it was updated.

9.8.3   Results

"Thing-speak" was the internet application used to interpret sensor-derived air quality data. A programming interface for the internet of things that uses the html protocol and can be accessed via the online or purpose of contributing network is known as Thing-speak. In addition, a wide range of embedded devices and web services can be accessed through it. As a result, sensor recording apps can be developed that are easily updated.

Before the sensor began measuring the sample aerosol, there was just a little amount of pollutant present. However, the air quality immediately plummeted from 0 to 100 ppm when the sensor identified the aerosol. There was a considerable decrease in the sample aerosol level in the air after many measurements were taken on different days.

On the 28th of February, the amount of dust in the air was at its lowest, but it rose steadily over the next few days. The air quality changed gradually on certain days, while being constant on others. The amount of dust in the air is affected by a wide range of circumstances.

When compared to the previous pollutants, the air quality was noticeably worse. After only a few days of taking data, it is clear that the air quality has deteriorated rapidly. This is the case due to the fact that gases are extremely polluting sources of air.

Monitors the quality of the air in a given area in real time and shows the measured air quality on a large LCD screen. The method helps people become more aware of the air they breathe on a regular basis. Using this air quality monitor, you will be able to see current air quality conditions in real time.

9.9 Irrigation Automation

India is mostly a farming nation. For the vast majority of Indian families, agriculture is their primary source of income. It is critical to the growth of an agricultural country like Australia. Agriculture in India accounts for around 16% of overall GDP and 10% of total exports. Agriculture relies heavily on water as a resource. While irrigation is one means of supplying water, there will be water wasted in other circumstances [30].

Farmers typically cultivate a wide range of crops on vast tracts of land. The entire farmland cannot be monitored by a single person at all times. It is possible that a certain area of land may receive more water than usual, causing flooding, or that it will receive significantly less water than usual, leaving the soil parched. Crops can be harmed in either scenario, and farmers may incur losses as a result [31].

Remote water supply management is made easier with the IoT Irrigation Monitoring and Control system. This system utilises the Internet of Things (IoT). A Wi-Fi module can be used to connect the system to the internet. The control signals are sent via an Arduino Uno board, which is linked to the internet.

Two things is shown on the website:

a. Status of the motor
b. Moisture content

Moisture sensors in the circuit maintain tabs on the soil's moisture content, and the website's "Moisture level" is updated accordingly. The user would be able to monitor the moisture level from a distance and adjust the water delivery accordingly. To switch on or off the "water pump," all the user has to do is toggle the "Motor status" on or off, as desired. As a result, the "soil moisture" is tracked and the "water supply" is managed simply by flipping the "Motor status" switch. As a result, the user will not have to be concerned about "water-logging" or "drought" damaging his crops or plants. If you have a tiny garden and cannot be always there, this device can help you monitor "soil moisture" and maintain correct water supply even from a distance without having to be physically present in your garden.

References

1. https://light-it.net/blog/9-prominent-benefits-of-iot-for-business/.
2. https://www.mobinius.com/blogs/iot-in-automation-benefits-impact.
3. Harrison, C., Eckman, B., Hamilton, R., Hartswick, P., Kalagnanam, J., Paraszczak, J., Williams, P., Foundations for smarter cities. *IBM J. Res. Dev.*, 54, 4, 1–16, 2010.
4. Jin, J., Gubbi, J., Marusic, S., Palaniswami, M., An information framework for creating a smart city through internet of things. *IEEE Internet Things J.*, 1, 2, 112–121, 2014.
5. Badis Hammi *et al.* Internet of Things (IoT) Technologies for Smart Cities", *IET Journals* The Institution of Engineering and Technology, 1-14, 2015.
6. Syed, A.S. *et al.*, IoT in smart cities: A survey of technologies, practices and challenges. *Smart Cities*, MDPI, 4, 2, 429–475, 2021.
7. https://bluespeedav.com/blog/item/7-greatest-advantages-of-smart-home-automation.
8. Majeed, R. *et al.*, An intelligent, secure, and smart home automation system. *Sci. Program.*, 2020, Article ID 4579291, 14, 2020.
9. Chong, G., Zhihao, L., Yifeng, Y., The research and implement of smart home system based on Internet of Things, in: *Proceedings of the 2011 International Conference on Electronics, Communications and Control (ICECC)*, September 2011, IEEE, Ningbo, China, pp. 2944–2947.
10. Haq, Z.U., Khan, G.F., Hussain, T., A comprehensive analysis of XML and JSON web technologies. *New Dev. Circuits Syst. Signal Process. Commun. Comput.*, 102–109, 2013.
11. Vapnik, V.N., *Nature of statistical learning theory*, Springer-Verlag, New York, NY, USA, 1995.
12. Han, J., Pei, J., Kamber, M., *Data mining: Concepts and techniques*, Morgan Kaufmann, Burlington, MA, USA, 2011.
13. Nakamoto, S., *Bitcoin: A peer-to-peer electronic cash system*, Triumph Books, Chicago, IL, USA, 2008.
14. Christidis, K. and Devetsikiotis, M., Blockchains and smart contracts for the Internet of Things. *IEEE Access*, 4, 2292–2303, 2016.
15. IBEF Presentation, [Online]. Available: https://www.ibef.org/download/Healthcare-February-2018.pdf.
16. PricewaterhouseCoopers, *Reimagining the possible in the Indian healthcare ecosystem with emerging technologies*, PwC, [Online]. Available: https://www.pwc.in/industries/healthcare/reimagining-the-possible-in-the-indian-healthcare-ecosystemwith-emerging-technologies.html.
17. Naresh, V.S. *et al.*, Internet of Things in healthcare: Architecture, applications, challenges, and solutions. *Int. J. Comput. Syst. Sci. Eng.*, 35, 6, 411–421, 2020.
18. Shanin, F. *et al.*, Portable and centralised E-health record system for patient monitoring using Internet of Things (IoT). Presented at the *International*

CET Conference on Control, Communication, and Computing (IC4), pp. 165–170, 2018.

19. Swaroop, K.N., Narendra Swaroop, K., Chandu, K., Gorrepotu, R., Deb, S., A health monitoring system for vital signs using IoT. *Internet Things*, 5, 116–129, 2019.

20. Rathore, M.M., Ahmad, A., Paul, A., Wan, J., Zhang, D., Realtime medical emergency response system: Exploiting IoT and big data for public health. *J. Med. Syst.*, 40, 12, 283, Dec. 2016.

21. Rohokale, V.M., Prasad, N.R., Prasad, R., A cooperative Internet of Things (IoT) for rural healthcare monitoring and control. *2011 2nd International Conference on Wireless Communication, Vehicular Technology, Information Theory and Aerospace & Electronic Systems Technology (Wireless VITAE)*, 2011.

22. Onasanya, A. and Elshakankiri, M., Smart integrated IoT healthcare system for cancer care. *Wirel. Netw.*, 27, 6, 4297–4312, 2021.

23. Sood, S.K. and Mahajan, I., Wearable IoT sensor based healthcare system for identifying and controlling chikungunya virus. *Comput. Ind.*, 91, 33–44, 2017.

24. Abdelgawad, A., Yelamarthi, K., Khattab, A., IoT-based health monitoring system for active and assisted living, in: *Smart Objects and Technologies for Social Good*, pp. 11–20, 2017.

25. Yang, Z., Zhou, Q., Lei, L., Zheng, K., Xiang, W., An IoT-cloud based wearable ECG monitoring system for smart healthcare. *J. Med. Syst.*, 40, 12, 286, Dec. 2016.

26. https://www.ixon.cloud/knowledge-hub/7-practical-applications-of-IOT-in-industrial-automation.

27. https://nexusintegra.io/7-industrial-iot-applications/.

28. https://www.iotchallengekeysight.com/2019/entries/smart-land/211-0515-025039-real-time-air-quality-monitoring-system-based-on-iot.

29. Okokpujie, K. *et al.*, A smart air pollution monitoring system. *Int. J. Civ. Eng. Technol.*, 9, 9, 799–809, September 2018.

30. https://nevonprojects.com/iot-irrigation-monitoring-controller-system/.

31. Hwang, J., Shin, C., Yoe, H., Study on an agricultural environment monitoring server system using wireless sensor networks, *Sensors*, 10, 12, 11189–11211, 2010.

10

Integration of IoT in Energy Management

Ganesh Angappan¹*, Santhosh Sivaraj², Premkumar Bhuvaneshwaran³, Mugilan Thanigachalam⁴, Sarath Sekar³ and Rajasekar Rathanasamy¹

¹Department of Mechanical Engineering, Kongu Engineering College, Perundurai, Erode, Tamil Nadu, India
²Department of Robotics and Automation, Easwari Engineering College, Ramapuram, Chennai, Tamil Nadu, India
³Department of Food Technology, Kongu Engineering College, Perundurai, Erode, Tamil Nadu, India
⁴Department of Mechanical Engineering, Government College of Technology, Coimbatore, Tamil Nadu, India

Abstract

Digital technologies in recent years have conquered the energy-based industries for achieving better business opportunities through adoption of certain business strategies. The main advantage of Internet of things (IoT) in energy sector was to interconnect the automate and monitor regularly through software interface. So that, the defect or error, which occurred during production process, can be easily located and resolved. Through online condition monitoring, the performance of the machine can be assessed with the application of certain algorithms. In energy industries, IoT made huge revolution in energy management, automation and distribution. IoT facilitates huge profit to investors for improving the workflow with reference to the data based on demand and usage of product. The devices with IoT would communicate each other and perform many tasks without any human intervention. Thus reduces human effort and minimizes the overall cost. Some of the major drawbacks of IoT in energy sectors are security issues and development of tasks for performing more cycle of works were tedious one. The importance of energy usage and their management in various energy sectors are discussed in this chapter.

Keywords: Energy management, mechanical vibration, smart grid, Industry 4.0, transportation of energy, internet of things

*Corresponding author: ganesh.mech@kongu.edu

R. Rajasekar, C. Moganapriya, P. Sathish Kumar and M. Harikrishna Kumar (eds.) Integration of Mechanical and Manufacturing Engineering with IoT: A Digital Transformation, (271–304) © 2023 Scrivener Publishing LLC

10.1   Introduction

Industrialization can be viewed in four stages, the initial one was the identification of alternate energy sources. However, increased steam generation provided the mass development of power plants utilizing the newly discovered alternate fuels [1]. The next stage provided for the increased industrial output especially in steel and iron manufacturing due to the invention of the assembly line process, this eventually lead to the formation of new private industries and enterprises that revolutionized the modern industrial process [2]. The following stage of the industrial revolution saw the rising of computers, communication services, and automation in the industrial processes [3]. And the last and most recent is the advent of artificial intelligence, discoveries in communication services, robotics, and the Internet of Things [4–6].

Industrial Internet is an emerging technology that aids global networks to communicate with each other, because many sectors are working on IoT, it will play an important part in future innovations. Communication is key for the industries and vendors as it provides better customer service, business intelligence and meets the vendor's needs by providing better analytics. As a result, industries are quickly embracing this technology to meet the demands of creative technical solutions and increased competition [7, 8]. The Internet of Things (IoT) was first proposed for connecting both radiofrequency identifiers and the Internet [9]. The Internet of Things is currently maintaining capillary networking infrastructure, which represents a sizable count of wireless devices that connect with the Internet [10, 11].

The Internet of Things links devices, the public, information, and developments by effortlessly connecting them. As a result, by gathering and processing massive amounts of data, IoT can aid in making many processes more quantifiable and observable [12]. This new system would immensely help people in their daily life with the likes of medicine, agriculture, banking, waste management, construction, etc. by automating the decision making in real time [13]. Management of traffic, health services, smart planning, and environmental planning are some of the applications of IoT [14].

Vibrating sources produce energy based on their frequency of vibration, mass, and amplitude, for example, wind energy is created by the movement of blades concerning the wind movement, speed, and direction as same as hydroelectricity produced by the movement, speed, and mass flow of the water. The replacement of WSN batteries is challenging due to environmental constraints, aside from that, batteries are incredibly expensive and pollute the environment. As a result, researchers are urged to explore different sources of energy: aeroelastic, wind energy, mechanical vibration, solar energy, thermal energy, sound energy, and radiofrequency energy, which can be converted into sustainable electric energy, specifically for WSN (commonly identified as clean, long-term sustainable and green energy sources) [15, 16].

According to the United Nations Sustainable Development Goals agenda, energy efficacy was only the primary aspects of sustainable development [17]. Furthermore, energy efficacy has long-standing economic benefits through minimizing the cost of raw material for fuel, increased clean-energy generation, and lowering energy-related pollutants. Real-time monitoring of the energy supply chain is critical for improving energy efficiency and optimizing energy management [18]. Figure 10.1 depicts the primary sections of the energy supply chain.

The ideal energy harvester would be able to deliver consistent energy, which is a difficult task. Management and energy storage can help to optimize this. IoT has been widely adopted around the world for several years, for example, by the end of 2020, 50 billion devices approximately gets interfaced with the Internet [20]. This work examines the usage of the IoT in managing the energy sector mainly in areas of generation, energy supply to end users, energy distribution and supply/demand. Hence, the foremost objective of this study is to give the policymakers, experts, and leaders the prospects and challenges in implementing the IoT in the energy sector.

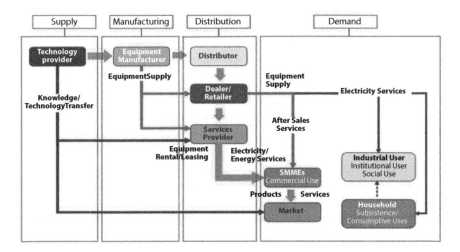

Figure 10.1 Energy supply chain [19].

10.2   Energy Management Integration with IoT in Industry 4.0

Reduced production costs, as previously stated, are critical for manufacturers to remain competitive in the fourth industrial revolution. One method to do so is to develop a robust energy management plan for Industry 4.0. Although it is critical to building a strong industrial ecosystem that provides a significant portion of national GDP, there has been a decline in industry's overall contributed value and social impact in recent decades, while unemployment rates have risen dramatically. One of the numerous variables that must be optimized for a company to remain competitive is energy efficiency, which is often overlooked in most businesses. Nowadays, advanced manufacturing or Lean Production principles are used by manufacturing systems all over the world, in which minimal resources are used to maximize business value.

For energy utilization, the same principles can and should be applied. From this standpoint, it is all about being efficient with how energy is used in the manufacturing system, reducing consumption whenever possible, and transferring enough knowledge to key individuals across the organization to deliver the same value with less energy. In Industry 4.0 energy management and efficiency, there is more than simply a requirement for energy availability, dependability, quality, and manageability. This is due to a combination of environmental issues, cost challenges, legislation, and other considerations, as seen by the increased focus on proactive energy consumption capacities (which is partly due to laws) and the integration of alternative energy sources.

Users shall possess more control and ability for regulating their energy usage, as energy becomes more digitally delivered with sensors, as well as more decentralized with energy coming from aeroelastic, radiofrequency, mechanical vibration, solar and thermal energy. There is no Industry 4.0 without comprehensive energy management at its core.

Whereas, traditional management of energy depend on effective delivery and use of process energy demands with the likes of cooling, heating, and electricity, the Internet of Things offers a variety of additional data streams to support energy management techniques. Due to their better expertise with the use of sensors and automation, manufacturing systems may not be that quick in implementing certain technologies as the consumer sector. This is primarily because maintaining global competitiveness and technological advancements is a key driver for the digital transformation of manufacturing systems, which forces the arrangement of production and wider business processes through tools that provide new business model possibilities.

Furthermore, a new breed of energy efficiency software tools allows two types of industrial energy management. An open-loop system is where the variables are entered manually for optimization, and closed loops are where the optimized values are set directly by the system. The open-loop model can save 3% to 8% of energy, while closed-loop applications can save 6% to 15% [21].

In light of the foregoing, the first step in implementing energy management in Industry 4.0 is to have a better understanding of energy flows and consumption, from which it will be possible to determine which consumers and consumptions are unnecessary or excessive. Then, to create a favorable climate for the implementation of continuous improvement strategies to manage action plans using continuous improvement methodology, appropriate indicators and real-time monitoring systems might be built [22].

It is crucial to note, however, that manufacturing operations can be thought of as being made of various levels from the perspective of manufacturing systems organization [23]. As a result, the choice of system abstraction levels is determined by the optimization research goal as well as the specific hierarchy levels. As illustrated in Figure 10.2, each hierarchical level consisting of energy resource, energy conversion and energy transfer, energy storage and control and energy consumption.

Each level also has a variety of ways to influence energy efficiency, each with its own set of effects. Classically, an industry has different levels

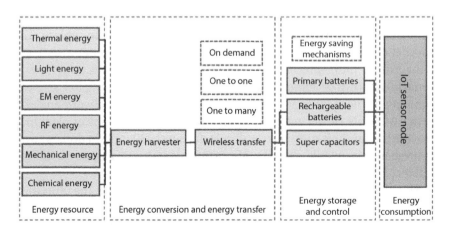

Figure 10.2 Energy harvesting system [24].

(departments, internal and external stakeholders) that have their own set of objectives of resource and energy utilization, this creates a disorderly system where holistically speaking the interactions, interdependency needs to be balanced to avoid trouble, local optimization, and modelling [25, 26].

10.3 IoT in Energy Sector

Roughly, 80% of total global energy is derived from fossil fuel, excessive mining and usage of fossil fuel created a negative impact on the health, environmental, and economic aspects with the likes of climate change, pollution, etc. Efficient use of energy is very important in reducing the impact of fossil fuel, thus new energy solutions that give the same service are required [27, 28]. The primary role of IoT in the energy sector is discussed here, this includes pollution monitoring, efficient fuel consumption, energy generation and maintenance, reduction in energy losses are some of the areas [18, 29].

10.3.1 Energy Generation

In the 1990s, industrial process automation, control and gathering data were become popular in the energy sector [30]. Early phases of IoT began to contribute to the power sector by reducing the hazard of output loss or blackout by monitoring and regulating equipment and operations. The main challenges of outdated power plants are consistency, output performance, effects on environment, and difficulties in maintenance. The working life of equipments in power sector, as well as inadequate maintenance issues, can result in significant energy losses and unreliability. Assets can be over 40 years old, extremely valuable, and difficult to replace. The Internet of Things (IoT) can help with some of these issues in power plant management [30]. Internet-connected equipment can detect any operational breakdown or abnormal drop in energy efficiency using IoT sensors, signaling the need for repair. This improves the system's resiliency and efficiency while also lowering maintenance costs. IoT driven power plant may provide the minimisation of overall cost (around 230 million dollars) over its lifespan. Meanwhile, same sized existing plant can save up to 50 million dollars [17].

Many countries are supporting RESs as a way to minimize the utilisation of fossil energy source as primary energy source. Climate based renewable energy sources such as wind and solar energy provide the energy system with new hurdles described as "the intermittency challenge." Equalizing the energy generation with energy demand with VRE is difficult due to fluctuation in the supply, resulting in mismatch across time scales of demand. IoT technology provide flexibility in harmonizing energy generation, reducing the difficulties in installing VRE and might end up with larger clean energy integration shares and lower GHG emissions [31].

Additionally, by utilizing IoT, more effective use of energy attained with the application of data science and AI learning, which assist in determining an appropriate balance of various supply and demand technologies [30]. Artificial intelligence systems, for example, can balance a thermal power plant's output with local power generation sources, aggregation of PV panels with thermal power plant [32].

10.3.2   Smart Cities

The rate at which urbanization happening in the world is creating bigger problems including water, air, and noise pollution [33], accessing low-cost energy, and environmental problems. One of the most significant difficulties in this area was attempting to have cleaner, inexpensive, and consistent power supply. Advancements in the areas of digital technology as fuelled the application of IoT devices in existing smart cities [34]. Connecting the city with all the relevant factors like electricity distribution, waste management, and essential supplies with IoT will enable frequent data collection ensuring flexibility like for example, if found that a factory in the area is consuming lots of power then these devices can be used to ensure that adequate power is supplied to it and not reducing the load on the transmission line.

Different processes in a smart city, such as transfer of information, communication, intelligent identification, pollution monitoring, etc. may all be perfectly controlled with the use of IoT devices [35]. IoT can assist in continuous monitoring of everything in the required field. Sensors could be installed in construction zones, urban infrastructure, transportation, energy networks, and utilities. By constantly monitoring data collected from sensors, these linkages help assure an energy-efficient smart city. For example, by using IoT to monitor automobiles, street lighting may be regulated to make the most efficient use of electricity. Furthermore, authorities will have access to the acquired data and will be able to make more prediction on energy supply and energy demand.

10.3.3   Smart Grid

Smart grids utilize the most dependable and secured ICT technology to ensure optimum energy is generated, transmitted, and distributed to the end-user. It creates a non-directional supply of information by connecting multiple smart meters, which can be used for system optimization and efficient energy distribution [36]. Besides, IoT can be placed on isolated, risk zones and islands where continuous monitoring of energy is to do like servers, satellite stations, etc., in which case all the connected assets can interact with each other, which would eventually assure the perfect energy distribution at all times.

In terms of the collective impact of smart grids, a smart city equipped with IoT-based smart grids, different sections of the city can be connected [36]. The smart grid can inform operators through smart appliances before any significant problem arises during collaborative communication across diverse sectors [29, 37].

The real-time monitoring of energy would ensure that the pricing would be based on the usage of energy. Generation and consumption of energy can be effortlessly optimized and managed by far-sighted plans using data collected from smart grid components. Other examples of IoT applications include reduction in transmission losses in T&D networks by active voltage control or reducing non-technical losses utilizing a network of smart meters [30].

10.3.4   Smart Buildings

Consumption of energy in smart city can be classified into three categories: residential, industrial (stores, offices, and schools), and transportation. Lighting, equipment, hot water supply for homes, cooking, refrigeration, heating, air-con, and ventilations are all examples of local energy usage in the housing sector (HVAC). In most buildings, HVAC energy usage accounts for half of the total energy use [38]. As a result, HVAC system control is critical for lowering electricity use. IoT devices with current technology surely capable of minimizing energy losses in HVAC systems. Unoccupied spaces, for example, can be identified using wireless thermostats. Once an uninhabited zone has been identified, various energy-saving measures can be performed. HVAC systems, for example, can limit operation in empty zones, resulting in significant decreases in energy consumption and losses. Figure 10.3 illustrates the consumption shares of residential energy utilisation.

Lighting system energy losses can also be managed using IoT. Customers will be notified when their energy use exceeds a certain threshold, for example, by using IoT-based lighting systems. In addition, the load can be shifted from higher levels to those of lower ones to make the loads even and effective. This contributes significantly to the effectual use of electrical

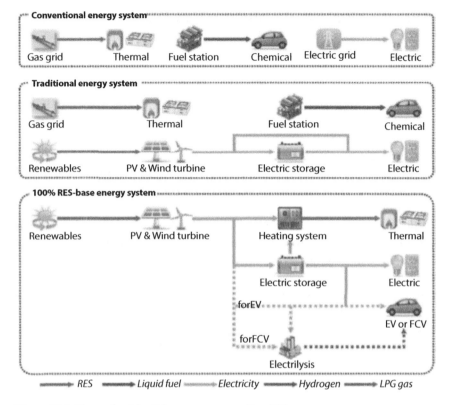

Figure 10.3 Share of residential energy consumption [19].

energy along with sequential reduction in emission of harmful greenhouse gases [34]. Demand response will be more nimble and versatile as a result of the IoT, as well as monitoring and demand-side management.

10.3.5 IoT in the Energy Industry

In energy sector, IoT was adopted for creating inter-linked autonomous system that reduces energy usage while increasing productivity. In a normal industry, quite a lot of energy is used in the production of the end product while controlling its quality, also at every step human presence is required to oversee every process. However, in smart factories, adopting an agile and adaptable system helps to spot faults at the same time rather than simply monitoring items after manufacturing process. This in turn provides corrective measures to avoid inefficient manufacturing associated with energy waste. For attaining flawless monitoring of manufacturing process, IoT technology was highly recommended [39].

Adding to it when multiple sensors are placed throughout the industry at critical points would allow the staff to find irregularities in the consumption of energy, which can be rectified without physical presence.

As a result, every single component can be readily maintained, component failures can be repaired, and each component's energy consumption may be improved. This effectively results in energy losses in smart industries being reduced. Data processing is the most important component of a small factory, as it allows data for analysis in the cloud platform, in order to achieve more effective decisions in real-time [40]. The depreciation of machines and mechanical devices is a major challenge in factories when it comes to monitoring and maintaining production assets.

The optimal device size can be selected using a suitable IoT platform and tools to minimize the wear and tear of a component along with its associated maintenance cost. Furthermore, addressing of failures that result in energy loss is expected. Systems with IoT could offer an effective solution for the customer, manufacturer, and company collaboration. As a result, a specific product will be made in direct response to a customer's request. Only a limited quantity of items of various types will be made and kept, allowing for better energy management and efficient production [39].

10.3.6   Intelligent Transportation

Usage of personal transport in place of public transport has created a few problems like pollution, stress and traffic congestion. Linking the entire transport system together would help in smart transportation applications like traffic control and smart parking systems that use online mapping. Passengers who use transportation wisely might choose a more cost-effective choice that consists of a faster route with less time of travel and to save resources [35]. Consumers shall plan ahead of time and to manage schedules [41]. As a result, city journeys will take less time, and energy losses will be greatly minimized. This has the potential of reducing greenhouse gases and other air pollutants from transportation [34].

The IoT-based digitization of an energy system turns it in single direction to an integrated energy system, i.e., from energy generation source by means of mart grids to the domestic end users. Figure 10.4 depicts several components of a smart energy system that is integrated.

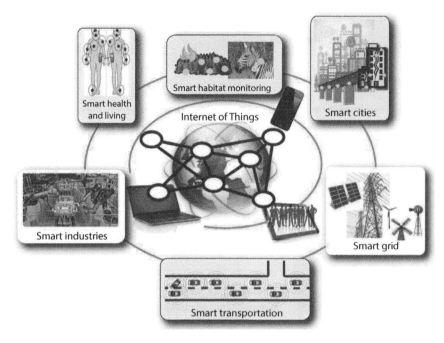

Figure 10.4 Integrated smart energy system [24].

10.4 Provocations in the IoT Applications

Apart from energy-saving benefit of IoT, implementing IoT especially in the energy business presents several difficulties need to be resolved. The above part highlights the issues that IoT-based energy systems confront, as well as the solutions that are currently available.

10.4.1 Energy Consumption

IoT platforms consist of several devices that are in constant communication for the transfer of information between them, to make this happen a lot of energy is required thus forming the first issue [42]. As a result, IoT systems' energy consumption remains a significant issue. Multiple techniques were been implemented for minimizing the power consumption of IoT devices such as making system to idle condition whenever not in working condition, etc. Figure 10.5 summarises the difficulties and current solutions in the energy sector for implementing IoT.

There has been a lot of research on how to build effective command transfer through signals using certain predefined procedures which permit

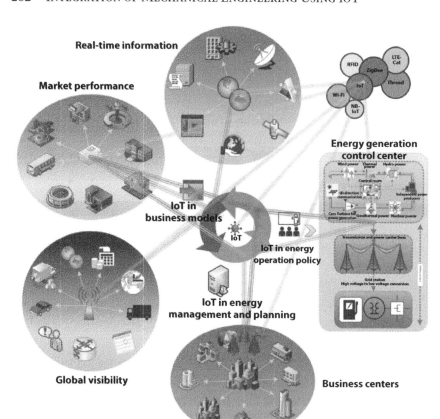

Figure 10.5 IoT employment in various sectors [43].

distributed computing systems to offer energy-efficient communications. As a possible solution, techniques of radiooptimization: modulation and cooperative communication were being offered. In addition to this, energy-efficient routing alternatives, say for example; grouped structures and different directional routing systems [44, 45]. In addition, Figure 10.6 represents the challenges in smart city energy management.

10.4.2   Subsystems and IoT Integration

Integration of IoT systems into energy system subsystems is a significant challenge. Because energy sector subsystems are unique, they utilize a variety of sensor as well as data connection methods. As a result, a need in strategy is required for efficient communication between the subsystems in IoT-enabled devices is necessary [47, 48]. Modelling an integrative approach

Figure 10.6 Challenges in IoT based smart city concept [46].

for the power system is a strategy for addressing the integration challenges while also taking into account the subsystem's IoT needs [47]. Another way for integrating the system and lowering synchronisation delay error across subsystems is to create co-simulation models for energy systems [49, 50].

User Privacy
When personal information is exchanged with an organization, individuals, or cooperative energy, the users have the right to privacy [51, 52]. As a result, accurate data on energy consumers, as well as the quantity and types of energy-consuming appliances, is impossible to get. Indeed, data obtained through the Internet of Things enables better policymaking that can affect energy production, supply and utilization [53]. Reduce the risk of violating consumers' confidentiality, energy providers should seek permission to use their data [54], ensuring that the data is not shared with third parties. Another option is to give power to the consumer about the usage of their personal information [55].

Security Challenge
The most important risk in this cloud-based world is the cyber threat, keeping user's data safe is the highest priority and this becomes difficult due to the interconnection of IoT devices with communication devices [56, 57].

The energy security challenge is characterized by these strands [58]. In addition, IoT-based energy systems are widely used in the energy industry to provide services across vast geographical areas. IoT systems are particularly vulnerable to assaults because of their widespread implementation. To address the problem, a study by Song *et al.* [59] investigates the possibility of an encryption mechanism to protect energy data from cyber attacks, additionally, a centralized control mechanism that enables a step level security at different stages of the system suggests the prevention of cyber security attacks [60].

IoT Standards

IoT devices connect a specific object to a huge quantity of other equipments, using a variety of technologies and protocols. A new difficulty arises from the discrepancy among different standards are used by IoT devices [61]. IoT-connected systems consist of two types of standards: data security and privacy network and communication protocols, data aggregation and regulatory standards. The adoption of standards in the IoT is hampered by difficulties such as unstructured data processing standards, security and privacy concerns, and legal requirements for data marketplaces [17]. To overcome the difficulty of standardization in IoT-based energy systems, define a systematic structure with a shared knowledge that all actors may access and use equally. Another possibility is for cooperating parties to develop information-sharing models and protocols for standardization. As a result, standards will be offered for free and to the general public [62].

Architecture Design

With the rising of smart networked devices and sensors, a wide range of technologies is used in IoT-enabled systems. IoT is expected to offer self-contained and ad hoc communications for any associated services at anytime, anywhere. Based on their application requirements, IoT systems are built with sophisticated, decentralized, and mobile characteristics [62]. A reference design, when considering the features and objectives of an IoT application, cannot be a one-size-fits-all solution for all of these applications. As a result, heterogeneous reference architectures that are open and follow standards are required for IoT systems. Users should not be restricted to fixed and end-to-end IoT communications by the architectures [63].

10.5   Energy Generation

Traditional types of energy such as light, sound, and wind are turned into electrical energy that may be utilized to power equipment in the process of

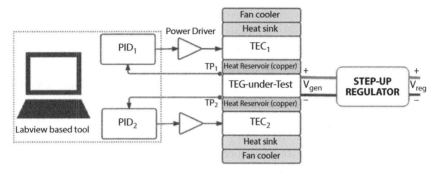

Figure 10.7 Workflow for thermal energy generation [66].

energy harvesting. Several approaches have been created throughout the years to use this principle for a variety of various sources [64]. Power generation acting an important part in driving IoT, harvesting system is a collection of subsystems that take into consideration of all available sources and efficiently convert them into usable electrical energy [64, 65]. Figure 10.7 depicts the process of thermal energy harvesting.

10.5.1   Conversion of Mechanical Energy

IoT requires a variety of sensors for healthcare, protection, condition monitoring, health management, and nature conservation. Batteries are a poor alternative for IoT due to several difficulties, including high costs, frequent maintenance, limited lifetime, and environmental concerns [67–69]. Some IoT devices require self-powering to function consistently throughout time. Since 2006, the Wang group has been developing nanogenerators that are possible to convert minute amounts of mechanical energy into electrical current [70]. The triboelectrification and piezoelectric effect were used to create the first nanogenerators.

Using the Maxwell equations as a starting point, investigation with the similarities and differences between triboelectrification energy harvesters, piezoelectric nanogenerators, and standard electromagnetic generators was made in 2017 [71]. Piezoelectric energy harvesters have been extensively explored, due to their high conversion efficiency and simple setup [72]. Using standard MEMS manufacturing techniques, creation of a multi-cantilever piezoelectric generator comes into the picture. A magnetically assisted stencil printing (MASP) process for fabricating ceramic devices with many layers were also developed using IoT technology [73]. Energy harvesting and other sensors can benefit from nanofibers composed of piezoelectric polymers [74]. Low-power electronic devices can be powered by energy recovered from unwanted mechanical vibration and abandoned heat [68, 75]. Vibration energy harvesting is the process of turning electrical energy by converting energy from the surrounding environment (VEH) [76].

It has been proposed to develop a piezoelectric impact-induced vibration cantilever energy system with a high-efficiency power management circuit and low power consumption. The harvester appears to be promising for converting low-frequency input created by cars travelling over speed bumps into high-frequency resonant output and gathering low-frequency excitement. Energy harvesting is most commonly used for micropower sources, although big harvesters can also be used for macro-power sources. Sung research group constructed piezoelectric cantilever beam-based power harvester used for a congested highway [77]. Over the previous three decades, piezoelectric energy harvesters have made a significant contribution to MEMS, as seen in Figure 10.6 [78]. The inherent frequencies of 48 piezoelectric beams were tweaked using road vibrational frequency. Vibration energy could be used to power street lamps in the future, and extra sensors could be attached to them to detect accidents. Vibration, motion capture, and acoustic energy can all be converted into electricity via piezoelectric energy harvesting (PEH) [79].

Before putting the Harvester to use in the real world, it must be reliable. The shock-absorbing construction increased the shock reliability of MEMS electrostatic vibration energy harvesters [80]. The harvester's base was a silicon mass-spring mechanism. The impact between spring anchors led the harvester to collapse under shock excitation. Shock-absorbing bumpers were established to improve impact resistance and move the shock's effect away from critical places. In 2009, Wang's group used a variety of mechanical stress and well-built periodic electric fields to test the piezoelectric actuator's dependability [81].

The findings revealed that as the load cycles were increased, mechanical strain and charge density decreased in a monotonic manner. Preload stress influenced the rate of deterioration [82]. In addition to this, the fatigue failure of PEH microfiber composites (MFC) due to excessive stress or strain. The PEH under investigation consisted of a cantilever beam with an MFC patch at one end. The Internet of Things, as well as compact self-power production and long-term energy supply, necessitated the development of triboelectric nanogenerators.

Methods of vibration diagnostic using IoT

Big data processing technologies and an artificial intelligence, multi-source sensing of data that originated from the IoT will enhance the durability of mechanical components, and bring down the labour cost of diagnosing mechanical faults.

Online Diagnostics methods using IoT
Identifying suspension problems and predicting maintenance online is proposed in the free-form model. The OEM research centre, cloud server, data analytics, and vehicle unit are included in its systems. Accelerometers are mounted on suspension systems, and the vehicle is equipped with GPS and communications. Cloud communication is possible with V2X for ECUs. OEMs can quote suspension systems vendors on this requirement for direct coupling with studs for good results. Each user and vehicle model could have a cloud-based knowledge database that can be used to evaluate similar cases so that suspension systems can be designed better and also various machine learning algorithms can be applied to different fault types [83].

Monitors the vibration of induction motors using an IoT platform. A 27 mm piezoelectric sensor is used to collect vibration data from the motor. Using two 27 mm piezo electrics in series, each IoT node is self-powered by vibration. Three different conditions have been examined in order to gather vibration data at different positions: bottom, top, and coupling. In each circumstance, the induction motor speed is increased or decreased. Our findings indicate that the suggested method is capable of recording motor vibrations with clarity and precision. Mobile phones are able to display results. There is an average delay of about 1 second between each node and the cloud when sending the recorded vibration [84]. Figure 10.8 indicates the working mechanism of self-powered vibration sensor device.

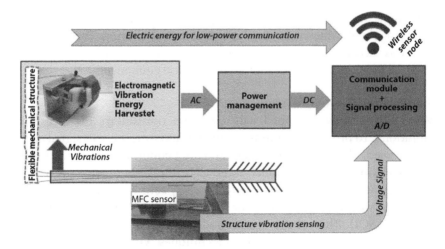

Figure 10.8 Self-powered vibration sensor [85].

Multi-source sensing data fusion for IoT diagnosis
Some unique scenarios can be identified using a neural network and DSS evidence theory, as well as some advanced data pre-processing algorithms. Artificial feature extraction for mechanical failure diagnosis in more complicated scenarios can be efficiently avoided using deep learning-based correlation fusion methods. Additionally, applying deep learning-based combination algorithms necessitated a good collection of hyper parameters. Some cases could benefit from a combination of heuristic algorithms for optimizing the network structure and parameters, like evolutionary algorithms, particle swarm optimization, and genetic algorithms.

Present study reveals that incorporating the advantages of different fusion algorithms into a more intelligent fusion model poses challenges as well as certain possibilities for diagnosing mechanical failure [86]. In air conditioners, installing the IoT sensors can be difficult, and designing an anomaly detection and diagnosis system might be complicated. Seasonal fluctuations and environmental change can complicate the identification and diagnosis of anomalies in air conditioners, in addition to loud environments and economic restrictions. Thus, an efficient and inexpensive anomaly detection system can be achieved by carefully coordinating data collection and diagnosis methods, as well as seasonally tuning the diagnosis model [87].

Methods of Vibration Monitoring In Electro-Mechanical Devices
In mechanical and electrical equipments, vibration is a significant characteristic that determines system reliability, efficiency, and durability. Performing vibration analysis on machine parts gives information about anomalous conditions of the machine [88]. It is important to measure vibrations before analyzing them, which is essential in condition monitoring. There are many methods such as wireless sensor network, algorithms and sensors devices are used for monitoring the vibration in mechanical equipment [89].

Subsynchronous control interactions (SSCI)
Wind turbines are equipped with vibration transducers for measuring vibrations. To process vibrations, frequency and time domain signal analyses are used. These vibration signals and fault conditions are characterized by correlating them. In parallel, three algorithms are used to process the input voltage signals, such as Prony, Eigen system Realization Algorithm (ERA), as well as Moving Window FFT in subsynchronous control interactions (SSCI). The voltage signal modes are calculated using ERA and Prony techniques as modal analysis techniques. Voltage input signals are calculated using Moving Window FFT [90].

Mechanical breakdowns in spinning equipment in pressure vessels and steam generators are occurring as a result of vibration. An on-line Condition Monitoring System has been used to monitor the main coolant pump shaft vibration as well as the general vibration of passive primary components. This approach is used to identify variations to reference signatures. Condition monitoring system with problem-oriented features must consider both active and passive failure characteristics [91].

Wireless vibration monitoring device with three axes
Vibration measurements of single-phase induction motors were performed by this system. Measurement of acceleration is accomplished using a triple axis accelerometer, ADXL335. To transfer vibration data, a ZigBee transceiver is used, a cost effective processor. Data is encrypted and decrypted at the receiver end using the Advanced Encryption Standard algorithm direct to safeguard against data eavesdropping. Lastly, the system's latency is measured. Mathematical filtering plays a crucial role in reducing noise in acceleration data, making it easier to obtain accurate vibration data. Double layers of protection are provided by zigbee and AES [92].

MEMS accelerometer sensors based on three axes have been utilized to measure vibration. Vibration sensing range of this sensor is from 0.0147 g to 7.5 g (1 g is 9.81 m/s^2). Advanced microcontroller is used with the accelerometer sensor on Arduino-derived microcontroller boards. In the implemented system, sensors and vibration monitors communicate wirelessly using the ZigBee protocol. XBee RF modules were used to conduct the wireless communication. In order to develop the PC's user interface, data-logging and automation alarms, LabVIEW software was used [93].

Multi-channel Micro-Electro-Mechanical Systems (MEMS)-based Low-Power Wide-Area Network (LPWAN)
In comparison with the separate wired and WSN-based monitoring systems, the proposed multichannel LPWAN monitoring system can provide accurate data synchronous collections, manage huge amounts of data, as well as monitor data streams in a broad coverage area. LPWAN monitoring system's amplitude and frequency values are 3.78% and 6.99% different from those of the wired monitoring system. The suggested synchronous approach is useful for machine vibration measurements, since it allows the synchronization errors to be controlled within 5 *s once they are synchronized instead of increasing linearly and gradually with acquisition time [94]. Radiofrequency identification tags, surface acoustic wave and sensors are used to simulate wireless vibration monitoring of rotating machines. Motion effects are compensated for by using a multi-channel interrogation approach.

By enhancing the signal-to-noise ratio, the vibratory signatures can be distinguished from low-power high-frequency components, and discriminatory information may be extracted from rotating machine parts. Vibration measurements on rotating shafts are used to verify final feasibility with induction motors. Since the transponder system is wireless and fully passive, unlike existing solutions, it can be positioned precisely at the location of interest, while existing solutions require power, storage, and connectivity. This allows for more appropriate measurements of phenomena such as strain and vibration [95].

Microelectromechanical systems (MEMS) accelerometers are used to increase the accuracy of vibration measurements on rotating machinery. Because the sampling clock in wireless applications is time-based, the data cannot be tied to a rotational phase. The dynamic behaviour of the rotor is investigated by mounting a wireless sensor unit and inverse encoder at one end of it. A method is also explored for decoding sensor data into vertical and horizontal vibrations. According to the data, the horizontal vibration demonstrated the vibration peaks are similar as in the reference measurement. In contrast to the reference measurement, the observed vibrations will not manifest themselves as clearly; therefore, further research is needed to validate the method [96].

Numerous studies were also interested in harvesters for satellite structures because the captured energy might be utilized to power IoT devices that would relay the signal or data to the ground station. Reconnaissance satellites, which contain a large number of sensors, employ piezoelectric harvesters (as in Figure 10.9) for this purpose. As a result, energy harvesting in reconnaissance satellites for IoT power is an important and promising challenge [98]. Elahi, H. et al. suggested a model for reconnaissance satellites that is efficient and driven by micro-electro-mechanical devices [99].

10.5.2   Aeroelastic Energy Harvesting

Several research have concentrated in recent years, on the potential source of energy to power IoT devices instead of traditional devices, and converting vibrations induced from a fluid flow was found to be promising [100]. With the use of such devices, those vibrations can be easily converted into electrical energy that can fully power or aid in additional power for the IoT devices. In the realm of aerospace engineering, these harvesters can be utilized in a wide range of ways. Take, for instance a scenario where a spacecraft is away from sunlight for long period, in those cases these harvesters

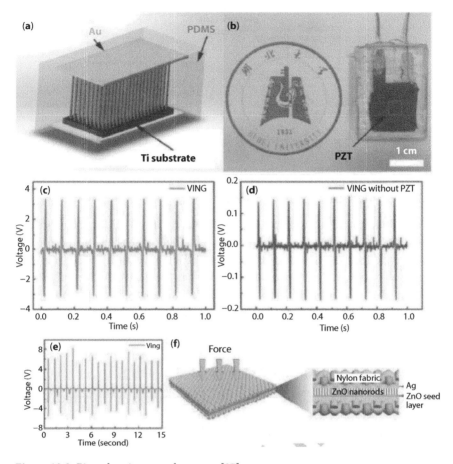

Figure 10.9 Piezoelectric energy harvester [97].

can be used to achieve the electrical energy needed to power the computers systems in the spacecraft [101].

Considering traditional battery backup would not be ideal because they have the disadvantages of being massive, having a very high operating expense, and, in some situations, such as on cruising altitude levels being hard to maintain [102]. In addition, the unstable aerodynamics system is strained in order to account for IoT harvesting energy [103]. In a subsonic wind tunnel with the harvester subjected in axial flow, numerical and experimental techniques were done. The critical velocity and bifurcation graph of the IoT harvesting phenomenon were also predicted [104]. Aeroelastic energy harvesting (as in Figure 10.10) based on galloping is also a potential subject for powering IoT devices [105].

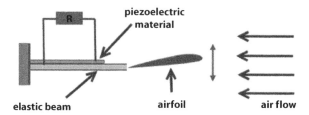

Figure 10.10 Wind energy to electric energy generation with the aid of piezoelectric effect [106].

10.5.3   Solar Energy Harvesting

Photovoltaic materials are used in the conversion of sun energy to electricity, many combinations of materials have been tried and tested for the past decade for improving the efficiency of the conversion [107]. The fundamental issue with photovoltaic cells is that they have higher efficiency outdoor than indoor [108]. Solar panels in the environment are used to power autonomous IoT devices, that can facilitate machine learning on the device [109].

10.5.4   Sound Energy Harvesting

Sound energy is abundant in the universe and can be easily extracted and converted into electrical energy, this can be done by the utilization of microsensors. Harvesting sound can result in a considerable loss of energy due to the intrinsic wavelength and disparity among mutual acoustic waves frequency. The majority of studies involving were to focus the majority of works utilizing sound energy harvesters [110]. Due to the device's limited bandwidth and inherent frequency, as well as the usual sound frequency mismatch, they had poor output performance. A triboelectric effect-based TENG is used to gather mechanical vibration energy [111, 112].

10.5.5   Wind Energy Harvesting

The climatic dependent renewable energy resource can be utilised effectively through proper automation and control. For example, remote sensors require extremely little power usage or an auto feature [113]. Self-sustaining fluctuations from a malleable piezoelectric membrane are used to harvest wind energy. Wind energy can be used in conjunction with hydro energy [114]. The vertical-axis and horizontal-axis windmills are two different types of windmills. The former denotes a condition in which

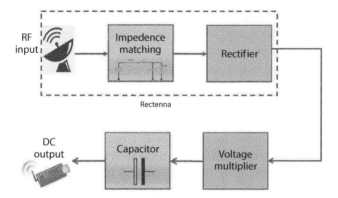

Figure 10.11 Workflow of RF energy harvester [24].

the wind direction is constantly changing, whereas the latter denotes a condition in which the wind direction remains constant.

10.5.6   Radiofrequency Energy Harvesting

Traditional energy gathering sources like thermal-electric impact, air, and sunlight are used in recharging IoT systems, though they are unstable because to huge fluctuations in energy levels. To mitigate this radiofrequency harvesting devices can be considered, the information can be carried concurrently with the energy production and processed simultaneously by the RF-based energy gathered signals [115]. This is for charging IoT devices particularly. Figure 10.11 depicts the workflow of radiofrequency energy harvesting system.

10.5.7   Thermal Energy

Temperature variation, thermal energy, and the major source "heat" The first is pyroelectric, and the second is thermoelectric, both of which are used to capture energy from a thermal source. The thermoelectric approach takes advantage of the Seebeck effect to directly transform temperature differences into useful energy. The thermoelectric harvesting technology can achieve a 5% to 8% harvesting efficiency. Thermoelectric devices are employed in a variety of applications on Earth and space. Many people have looked into using thermoelectric sources to harvest energy [116]. The possibility of converting waste heat to electricity for portable devices based on IoT is important, and this is done by pyroelectric energy devices that convert fluctuations in heat to electricity.

10.6 Conclusion

Thus, the implementation of IoT into various energy sectors such as mechanical, thermal, wind, sound, solar and aeroelastic sectors were represented in detail. The futuristic world will exceptionally rely on advanced technologies, which reduces the human effort with the achievement of very high performance and functionality. Occurrence of error, data security and breach will be the major concern in the IoT based technologies. However, error detection and resolution from time to time aspects should be performed. IoT as an emerging tool, can extend its approach in energy sector, as consumption becomes increasing in everyday situations. Therefore, energy management and automation are found to be necessary in present and future aspects. This might lead to explicit growth of business and economy of that particular field. With the aid of bigger communication technology, efforts in research were taken with the futuristic perspective made them unavoidable promising technology.

References

1. Stearns, P.N., *Reconceptualizing the industrial revolution*, J. Horn, L.N. Rosenband, M.R. Smith (Eds.), 362 pp. $24.00, The MIT Press, Cambridge, Mass, 2011, https://doi.org/10.1162/JINH_r_00261.
2. Mokyr, J. and Strotz, R.H., The second industrial revolution, 1870-1914. *Storia dell'economia Mondiale*, 21945, 1–16, 1998.
3. Jensen, M.C., The modern industrial revolution, exit, and the failure of internal control systems. *J. Finance*, 48, 3, 1993, https://doi.org/10.1111/j.1540-6261.1993.tb04022.x.
4. Henning, K., Recommendations for implementing the strategic initiative INDUSTRIE 4.0, *Final report of the Industrie*, Forschungsunion, acatech - Germany, 4, 1–82, 2013.
5. Witchalls, C. and Chambers, J., *The Internet of Things business index: A quiet revolution gathers pace*, The Economist Intelligence Unit, London, 58–66, 2013. Retrieved from http://www.economistinsights.com/analysis/internet-things-business-index.
6. Datta, S.K. and Bonnet, C., MEC and IoT based automatic agent reconfiguration in industry 4.0, in: *2018 IEEE International Conference on Advanced Networks and Telecommunications Systems (ANTS)*, IEEE, 2018, https://doi.org/10.1109/ANTS.2018.8710126.
7. Lee, I. and Lee, K., The Internet of Things (IoT): Applications, investments, and challenges for enterprises. *Bus. Horiz.*, 58, 4, 431–440, 2015, https://doi.org/10.1016/j.bushor.2015.03.008.

8. Gubbi, J., Buyya, R., Marusic, S., Palaniswami, M., Internet of Things (IoT): A vision, architectural elements, and future directions. *Future Gener. Comput. Syst.*, 29, 7, 1645–1660, 2013, https://doi.org/10.1016/j.future.2013.01.010.

9. Ashton, K., That 'Internet of Things' thing. *RFID J.*, 22, 7, 97–114, 2009.

10. Hersent, O., Boswarthick, D., Elloumi, O., *The Internet of Things: Key applications and protocols*, John Wiley & Sons, New Jersey, United States, 2011, https://doi.org/10.1002/9781119958352.

11. Palattella, M.R., Accettura, N., Vilajosana, X., Watteyne, T., Grieco, L.A., Boggia, G., Dohler, M., Standardized protocol stack for the Internet of (important) Things. *IEEE Commun. Surv. Tutorials*, 15, 3, 1389–1406, 2012, https://doi.org/10.1109/SURV.2012.111412.00158.

12. Shrouf, F., Ordieres, J., Miragliotta, G., Smart factories in industry 4.0: A review of the concept and of energy management approached in production based on the Internet of Things paradigm, in: *2014 IEEE International Conference on Industrial Engineering and Engineering Management*, IEEE, 2014, https://doi.org/10.1109/IEEM.2014.7058728.

13. Bandyopadhyay, D. and Sen, J., Internet of Things: Applications and challenges in technology and standardization. *Wirel. Pers. Commun.*, 58, 1, 49–69, 2011, https://doi.org/10.1007/s11277-011-0288-5.

14. Gilbert, J.M. and Balouchi, F., Comparison of energy harvesting systems for wireless sensor networks. *Int. J. Autom. Comput.*, 5, 4, 334–347, 2008, https://doi.org/10.1007/s11633-008-0334-2.

15. Botteron, C., Briand, D., Mishra, B., Tasselli, G., Janphuang, P., Haug, F.-J., Skrivervik, A., Lockhart, R., Robert, C., de Rooij, N.F., A low-cost UWB sensor node powered by a piezoelectric harvester or solar cells. *Sens. Actuators A: Phys.*, 239, 127–136, 2016, https://doi.org/10.1016/j.sna.2016.01.011.

16. Gusarov, B., Gusarova, E., Viala, B., Gimeno, L., Boisseau, S., Cugat, O., Vandelle, E., Louison, B., Thermal energy harvesting by piezoelectric PVDF polymer coupled with shape memory alloy. *Sens. Actuators A: Phys.*, 243, 175–181, 2016, https://doi.org/10.1016/j.sna.2016.03.026.

17. Hossein Motlagh, N., Mohammadrezaei, M., Hunt, J., Zakeri, B., Internet of Things (IoT) and the energy sector. *Energies*, 13, 2, 494, 2020, https://doi.org/10.3390/en13020494.

18. Tan, Y.S., Ng, Y.T., Low, J.S.C., Internet-of-Things enabled real-time monitoring of energy efficiency on manufacturing shop floors. *Proc. CIRP*, 61, 376–381, 2017, https://doi.org/10.1016/j.procir.2016.11.242.

19. Hasanuzzaman, M. and Kumar, L., Energy supply, in: *Energy for Sustainable Development*, Elsevier, Massachusetts, 2020, https://doi.org/10.1016/B978-0-12-814645-3.00004-3.

20. Evans, D., *The Internet of Things: How the next evolution of the internet is changing everything*. CISCO white paper, Singapore, 1–11, 2011.

21. Dwyer, B. and Bassa, J., Combining IoT, industry 4.0, and energy management suggests exciting future. *InTech Magazine*, Mar-Apr, 2018.

22. Medojevic, M., Medojevic, M., Cosic, I., Lazarevic, M., Dakic, D., Determination and analysis of energy efficiency potential in socks manufacturing system. *Annals of DAAAM & Proceedings*, vol. 28, 2017, https://doi.org/10.2507/28th.daaam.proceedings.082.

23. Duflou, J.R., Sutherland, J.W., Dornfeld, D., Herrmann, C., Jeswiet, J., Kara, S., Hauschild, M., Kellens, K., Towards energy and resource efficient manufacturing: A processes and systems approach. *CIRP Ann.*, 61, 2, 587–609, 2012, https://doi.org/10.1016/j.cirp.2012.05.002.

24. Zeadally, S., Shaikh, F.K., Talpur, A., Sheng, Q.Z., Design architectures for energy harvesting in the Internet of Things. *Renewable Sustainable Energy Rev.*, 128, 109901, 2020, https://doi.org/10.1016/j.rser.2020.109901.

25. Fang, K., Uhan, N., Zhao, F., Sutherland, J.W., A new shop scheduling approach in support of sustainable manufacturing, in: *Glocalized solutions for sustainability in manufacturing*, pp. 305–310, Springer, Berlin, Heidelberg, 2011, https://doi.org/10.1007/978-3-642-19692-8_53.

26. Kara, S. and Li, W., Unit process energy consumption models for material removal processes. *CIRP Ann.*, 60, 1, 37–40, 2011, https://doi.org/10.1016/j.cirp.2011.03.018.

27. Connolly, D., Lund, H., Mathiesen, B.V., Smart energy Europe: The technical and economic impact of one potential 100% renewable energy scenario for the European Union. *Renewable Sustainable Energy Rev.*, 60, 1634–1653, 2016, https://doi.org/10.1016/j.rser.2016.02.025.

28. Grubler, A., Wilson, C., Bento, N., Boza-Kiss, B., Krey, V., McCollum, D.L., Rao, N.D., Riahi, K., Rogelj, J., De Stercke, S., A low energy demand scenario for meeting the 1.5 C target and sustainable development goals without negative emission technologies. *Nat. Energy*, 3, 6, 515–527, 2018, https://doi.org/10.1038/s41560-018-0172-6.

29. Bhardwaj, A., *Leveraging the Internet of Things and analytics for smart energy management*, TATA Consultancy Services, Mumbai, India, 2015.

30. Ramamurthy, A. and Jain, P., *The Internet of Things in the power sector opportunities in Asia and the Pacific*, Asian Development Bank- Philippines, 8, 1–26, 2017, https://doi.org/10.22617/WPS178914-2.

31. Al-Ali, A., Internet of Things role in the renewable energy resources. *Energy Proc.*, 100, 34–38, 2016, https://doi.org/10.1016/j.egypro.2016.10.144.

32. Karnouskos, S., The cooperative Internet of Things enabled smart grid, in: *Proceedings of the 14th IEEE International Symposium on Consumer Electronics (ISCE2010)*, June 2010.

33. Lagerspetz, E., Motlagh, N.H., Zaidan, M.A., Fung, P.L., Mineraud, J., Varjonen, S., Siekkinen, M., Nurmi, P., Matsumi, Y., Tarkoma, S., Megasense: Feasibility of low-cost sensors for pollution hot-spot detection, in: *2019 IEEE 17th International Conference on Industrial Informatics (INDIN)*, IEEE, 2019, https://doi.org/10.1109/INDIN41052.2019.8971963.

34. Ejaz, W., Naeem, M., Shahid, A., Anpalagan, A., Jo, M., Efficient energy management for the Internet of Things in smart cities. *IEEE Commun. Mag.*, 55, 1, 84–91, 2017, https://doi.org/10.1109/MCOM.2017.1600218CM.

35. Mohanty, S.P., Choppali, U., Kougianos, E., Everything you wanted to know about smart cities: The Internet of Things is the backbone. *IEEE Consum. Electron. Mag.*, 5, 3, 60–70, 2016, https://doi.org/10.1109/MCE.2016.2556879.

36. Hossain, M., Madlool, N., Rahim, N., Selvaraj, J., Pandey, A., Khan, A.F., Role of smart grid in renewable energy: An overview. *Renewable Sustainable Energy Rev.*, 60, 1168–1184, 2016, https://doi.org/10.1016/j.rser.2015.09.098.

37. Karnouskos, S., Colombo, A.W., Lastra, J.L.M., Popescu, C., Towards the energy efficient future factory, in: *2009 7th IEEE International Conference on Industrial Informatics*, pp. 367–371, IEEE, 2009, https://doi.org/10.1109/INDIN.2009.5195832.

38. Vakiloroaya, V., Samali, B., Fakhar, A., Pishghadam, K., A review of different strategies for HVAC energy saving. *Energy Convers. Manage.*, 77, 738–754, 2014, https://doi.org/10.1016/j.enconman.2013.10.023.

39. Lee, C. and Zhang, S., Development of an industrial Internet of Things suite for smart factory towards re-industrialization in Hong Kong, in: *Proceedings of the 6th International Workshop of Advanced Manufacturing and Automation*, Manchester, UK, 2016.

40. Reinfurt, L., Falkenthal, M., Breitenbücher, U., Leymann, F., Applying IoT patterns to smart factory systems. *Advanced Summer School on Service Oriented Computing, Summer SOC*, vol. 66, 2017.

41. Arasteh, H., Hosseinnezhad, V., Loia, V., Tommasetti, A., Troisi, O., Shafie-khah, M., Siano, P., IoT-based smart cities: A survey, in: *2016 IEEE 16th International Conference on Environment and Electrical Engineering (EEEIC)*, IEEE, 2016, https://doi.org/10.1109/EEEIC.2016.7555867.

42. Kaur, N. and Sood, S.K., An energy-efficient architecture for the Internet of Things (IoT). *IEEE Syst. J.*, 11, 2, 796–805, 2015, https://doi.org/10.1109/JSYST.2015.2469676.

43. Shaikh, F.K., Zeadally, S., Exposito, E., Enabling technologies for green Internet of Things. *IEEE Syst. J.*, 11, 2, 983–994, 2015, https://doi.org/10.1109/JSYST.2015.2415194.

44. Lin, Y.-H., Chou, Z.-T., Yu, C.-W., Jan, R.-H., Optimal and maximized configurable power saving protocols for corona-based wireless sensor networks. *IEEE Trans. Mob. Comput.*, 14, 12, 2544–2559, 2015, https://doi.org/10.1109/TMC.2015.2404796.

45. Ahmad, T. and Zhang, D., Using the Internet of Things in smart energy systems and networks. *Sustainable Cities Soc.*, 68, 102783, 2021, https://doi.org/10.1016/j.scs.2021.102783.

46. Nižetić, S., Šolić, P., González-de, D. L.-d.-I., Patrono, L., Internet of Things (IoT): Opportunities, issues and challenges towards a smart and sustainable future. *J. Cleaner Prod.*, 274, 122877, 2020, https://doi.org/10.1016/j.jclepro.2020.122877.

47. Shakerighadi, B., Anvari-Moghaddam, A., Vasquez, J.C., Guerrero, J.M., Internet of Things for modern energy systems: State-of-the-art, challenges, and open issues. *Energies*, 11, 5, 1252, 2018, https://doi.org/10.3390/en11051252.

48. Anjana, K. and Shaji, R., A review on the features and technologies for energy efficiency of smart grid. *Int. J. Energy Res.*, 42, 3, 936–952, 2018, https://doi.org/10.1002/er.3852.

49. Kounev, V., Tipper, D., Levesque, M., Grainger, B.M., Mcdermott, T., Reed, G.F., A microgrid co-simulation framework, in: *2015 Workshop on Modeling and Simulation of Cyber-Physical Energy Systems (MSCPES)*, IEEE, 2015, https://doi.org/10.1109/MSCPES.2015.7115398.

50. Wong, T.Y., Shum, C., Lau, W.H., Chung, S., Tsang, K.F., Tse, C., Modeling and co-simulation of IEC61850-based microgrid protection, in: *2016 IEEE International Conference on Smart Grid Communications (SmartGridComm)*, IEEE, 2016.

51. Porambage, P., Ylianttila, M., Schmitt, C., Kumar, P., Gurtov, A., Vasilakos, A.V., The quest for privacy in the Internet of Things. *IEEE Cloud Comput.*, 3, 2, 36–45, 2016, https://doi.org/10.1109/MCC.2016.28.

52. Chow, R., The last mile for IoT privacy. *IEEE Secur. Privacy*, 15, 6, 73–76, 2017, https://doi.org/10.1109/MSP.2017.4251118.

53. Jayaraman, P.P., Yang, X., Yavari, A., Georgakopoulos, D., Yi, X., Privacy preserving Internet of Things: From privacy techniques to a blueprint architecture and efficient implementation. *Future Gener. Comput. Syst.*, 76, 540–549, 2017, https://doi.org/10.1016/j.future.2017.03.001.

54. Roman, R., Najera, P., Lopez, J., Securing the Internet of Things. *Computer*, 44, 9, 51–58, 2011, https://doi.org/10.1109/MC.2011.291.

55. Fhom, H.S., Kuntze, N., Rudolph, C., Cupelli, M., Liu, J., Monti, A., A user-centric privacy manager for future energy systems, in: *2010 International Conference on Power System Technology*, IEEE, 2010.

56. Dorri, A., Kanhere, S.S., Jurdak, R., Gauravaram, P., Blockchain for IoT security and privacy: The case study of a smart home, in: *2017 IEEE International Conference on Pervasive Computing and Communications Workshops (PerCom Workshops)*, IEEE, 2017, https://doi.org/10.1109/PERCOMW.2017.7917634.

57. Poyner, I. and Sherratt, R., Privacy and security of consumer IoT devices for the pervasive monitoring of vulnerable people, in: *Living in the Internet of Things: Cybersecurity of the IoT-2018*, IET, London, 2018, https://doi.org/10.1049/cp.2018.0043.

58. Li, Z., Shahidehpour, M., Aminifar, F., Cybersecurity in distributed power systems. *Proc. IEEE*, 105, 7, 1367–1388, 2017, https://doi.org/10.1109/JPROC.2017.2687865.

59. Song, T., Li, R., Mei, B., Yu, J., Xing, X., Cheng, X., A privacy preserving communication protocol for IoT applications in smart homes. *IEEE Internet Things J.*, 4, 6, 1844–1852, 2017, https://doi.org/10.1109/JIOT.2017.2707489.

60. Roman, R. and Lopez, J., Security in the distributed Internet of Things, in: *International Conference on Trusted Systems*, Springer, 2012, https://doi.org/10.1007/978-3-642-35371-0_6.

61. Meddeb, A., Internet of Things standards: Who stands out from the crowd? *IEEE Commun. Mag.*, 54, 7, 40–47, 2016, https://doi.org/10.1109/MCOM.2016.7514162.

62. Chen, S., Xu, H., Liu, D., Hu, B., Wang, H., A vision of IoT: Applications, challenges, and opportunities with china perspective. *IEEE Internet Things J.*, 1, 4, 349–359, 2014, https://doi.org/10.1109/JIOT.2014.2337336.

63. Al-Qaseemi, S.A., Almulhim, H.A., Almulhim, M.F., Chaudhry, S.R., IoT architecture challenges and issues: Lack of standardization, in: *2016 Future Technologies Conference (FTC)*, IEEE, 2016, https://doi.org/10.1109/FTC.2016.7821686.

64. Gaudenzi, P., *Smart structures: Physical behaviour, mathematical modelling and applications*, John Wiley & Sons, New Jersey, United States, 2009, https://doi.org/10.1002/9780470682401.

65. Lu, C., Raghunathan, V., Roy, K., Micro-scale energy harvesting: A system design perspective, in: *2010 15th Asia and South Pacific Design Automation Conference (ASP-DAC)*, IEEE, 2010.

66. Pereira, R.I., Camboim, M.M., Villarim, A.W., Souza, C.P., Jucá, S.C., Carvalho, P.C., On harvesting residual thermal energy from photovoltaic module back surface. *AEU-Int. J. Electron. Commun.*, 111, 152878, 2019, https://doi.org/10.1016/j.aeue.2019.152878.

67. Butt, Z., Pasha, R.A., Qayyum, F., Anjum, Z., Ahmad, N., Elahi, H., Generation of electrical energy using lead zirconate titanate (PZT-5A) piezoelectric material: Analytical, numerical and experimental verifications. *J. Mech. Sci. Technol.*, 30, 8, 3553–3558, 2016, https://doi.org/10.1007/s12206-016-0715-3.

68. Swati, R., Wen, L., Elahi, H., Khan, A., Shad, S., Extended finite element method (XFEM) analysis of fiber reinforced composites for prediction of micro-crack propagation and delaminations in progressive damage: A review. *Microsyst. Technol.*, 25, 3, 747–763, 2019, https://doi.org/10.1007/s00542-018-4021-0.

69. Ali, A., Pasha, R.A., Elahi, H., Sheeraz, M.A., Bibi, S., Hassan, Z.U., Eugeni, M., Gaudenzi, P., Investigation of deformation in bimorph piezoelectric actuator: Analytical, numerical and experimental approach. *Integr. Ferroelectr.*, 201, 1, 94–109, 2019, https://doi.org/10.1080/10584587.2019.1668694.

70. Wang, Z.L. and Song, J., Piezoelectric nanogenerators based on zinc oxide nanowire arrays. *Science*, 312, 5771, 242–246, 2006, https://doi.org/10.1126/science.1124005.

71. Wang, Y., Yang, Y., Wang, Z.L., Triboelectric nanogenerators as flexible power sources. *NPJ Flex. Electron.*, 1, 1, 1–10, 2017, https://doi.org/10.1038/s41528-017-0007-8.

72. Lin, S., Lee, B., Wu, W., Lee, C., Multi-cantilever piezoelectric MEMS generator in energy harvesting, in: *2009 IEEE International Ultrasonics Symposium*, IEEE, 2009, https://doi.org/10.1109/ULTSYM.2009.5441451.

73. Medesi, A.J., Hagedorn, F., Schepperle, M., Megnin, C., Hanemann, T., The co-casting process: A new manufacturing process for ceramic multilayer devices. *Sens. Actuators A: Phys.*, 251, 266–275, 2016, https://doi.org/10.1016/j.sna.2016.07.033.

74. Persano, L., Dagdeviren, C., Su, Y., Zhang, Y., Girardo, S., Pisignano, D., Huang, Y., Rogers, J.A., High performance piezoelectric devices based on aligned arrays of nanofibers of poly (vinylidenefluoride-co-trifluoroethylene). *Nat. Commun.*, 4, 1, 1–10, 2013, https://doi.org/10.1038/ncomms2639.

75. Swati, R., Elahi, H., Wen, L., Khan, A., Shad, S., Mughal, M.R., Investigation of tensile and in-plane shear properties of carbon fiber reinforced composites with and without piezoelectric patches for micro-crack propagation using extended finite element method. *Microsyst. Technol.*, 25, 6, 2361–2370, 2019, https://doi.org/10.1007/s00542-018-4120-y.

76. Chen, N., Jung, H.J., Jabbar, H., Sung, T.H., Wei, T., A piezoelectric impact-induced vibration cantilever energy harvester from speed bump with a low-power power management circuit. *Sens. Actuators A: Phys.*, 254, 134–144, 2017, https://doi.org/10.1016/j.sna.2016.12.006.

77. Song, Y., Yang, C.H., Hong, S.K., Hwang, S.J., Kim, J.H., Choi, J.Y., Ryu, S.K., Sung, T.H., Road energy harvester designed as a macro-power source using the piezoelectric effect. *Int. J. Hydrogen Energy*, 41, 29, 12563–12568, 2016, https://doi.org/10.1016/j.ijhydene.2016.04.149.

78. Elahi, H., Eugeni, M., Gaudenzi, P., A review on mechanisms for piezoelectric-based energy harvesters. *Energies*, 11, 7, 1850, 2018, https://doi.org/10.3390/en11071850.

79. Ilyas, M.A. and Swingler, J., Piezoelectric energy harvesting from raindrop impacts. *Energy*, 90, 796–806, 2015, https://doi.org/10.1016/j.energy.2015.07.114.

80. Fujita, T., Renaud, M., Goedbloed, M., de Nooijer, C., Altena, G., Elfrink, R., van Schaijk, R., Reliability improvement of vibration energy harvester with shock absorbing structures. *Proc. Eng.*, 87, 1206–1209, 2014, 2014, https://doi.org/10.1016/j.proeng.2014.11.384.

81. Wang, H., Wereszczak, A.A., Lin, H.-T., Fatigue response of a PZT multilayer actuator under high-field electric cycling with mechanical preload. *J. Appl. Phys.*, 105, 1, 014112, 2009, https://doi.org/10.1063/1.3065097.

82. Upadrashta, D. and Yang, Y., Experimental investigation of performance reliability of macro fiber composite for piezoelectric energy harvesting applications. *Sens. Actuators A: Phys.*, 244, 223–232, 2016, https://doi.org/10.1016/j.sna.2016.04.043.

83. Kokane, P. and Sivakumar, P.B., Online model for suspension faults diagnostics using IoT and analytics, in: *International Conference on Advanced*

Computing Networking and Informatics, Springer, 2019, https://doi.org/10.1007/978-981-13-2673-8_17.

84. Firmansah, A., Mufti, N., Affandi, A.N., Zaeni, I.A.E., Self-powered IoT based vibration monitoring of induction motor for diagnostic and prediction failure, in: *IOP Conference Series: Materials Science and Engineering*, IOP Publishing, 2019, https://doi.org/10.1088/1757-899X/588/1/012016.

85. Rubes, O., Chalupa, J., Ksica, F., Hadas, Z., Development and experimental validation of self-powered wireless vibration sensor node using vibration energy harvester. *Mech. Syst. Sig. Process.*, 160, 107891, 2021, https://doi.org/10.1016/j.ymssp.2021.107890.

86. Huang, M., Liu, Z., Tao, Y., Mechanical fault diagnosis and prediction in IoT based on multi-source sensing data fusion. *Simul. Modell. Pract. Theory*, 102, 101981, 2020, https://doi.org/10.1016/j.simpat.2019.101981.

87. Hirata, T., Yoshida, K., Koido, K., Takahashi, S., Anomaly detection in air conditioners using IoT technologies, in: *2021 IEEE 45th Annual Computers, Software, and Applications Conference (COMPSAC)*, IEEE, 2021, https://doi.org/10.1109/COMPSAC51774.2021.00231.

88. Jung, D., Zhang, Z., Winslett, M., Vibration analysis for IoT enabled predictive maintenance, in: *2017 IEEE 33rd International Conference on Data Engineering (ICDE)*, IEEE, 2017, https://doi.org/10.1109/ICDE.2017.170.

89. Iannacci, J., Microsystem based Energy Harvesting (EH-MEMS): Powering pervasivity of the Internet of Things (IoT)-A review with focus on mechanical vibrations. *J. King Saud Univ.-Sci.*, 31, 1, 66–74, 2019, https://doi.org/10.1016/j.jksus.2017.05.019.

90. Zhao, L., Zhou, Y., Matsuo, I., Korkua, S.K., Lee, W.-J., The design of a holistic IoT-based monitoring system for a wind turbine, in: *2019 IEEE/IAS 55th Industrial and Commercial Power Systems Technical Conference (I&CPS)*, IEEE, 2019, https://doi.org/10.1109/ICPS.2019.8733375.

91. Van Niekerk, F. and Sunder, R., COMOS-an online system for problemorientated vibration monitoring. *Prog. Nucl. Energy*, 21, 155–171, 1988, https://doi.org/10.1016/0149-1970(88)90031-5.

92. Hossain, N.I., Reza, S., Ali, M., VibNet: Application of wireless sensor network for vibration monitoring using ARM, in: *2019 International Conference on Robotics, Electrical and Signal Processing Techniques (ICREST)*, IEEE, 2019, https://doi.org/10.1109/ICREST.2019.8644495.

93. Upadhye, M.Y., Borole, P., Sharma, A.K., Real-time wireless vibration monitoring system using LabVIEW, in: *2015 International Conference on Industrial Instrumentation and Control (ICIC)*, IEEE, 2015, https://doi.org/10.1109/IIC.2015.7150876.

94. Gao, S., Zhang, X., Du, C., Ji, Q., A multichannel low-power wide-area network with high-accuracy synchronization ability for machine vibration monitoring. *IEEE Internet Things J.*, 6, 3, 5040–5047, 2019, https://doi.org/10.1109/JIOT.2019.2895158.

95. Caldero, P. and Zoeke, D., Multi-channel real-time condition monitoring system based on wideband vibration analysis of motor shafts using SAW RFID tags coupled with sensors. *Sensors*, 19, 24, 5398, 2019, https://doi.org/10.3390/s19245398.

96. Koene, I., Viitala, R., Kuosmanen, P., Vibration monitoring of a large rotor utilizing Internet of Things based on-shaft MEMS accelerometer with inverse encoder, in: *12th International Conference on Vibrations in Rotating Machinery*, CRC Press, 2020.

97. Ali, F., Raza, W., Li, X., Gul, H., Kim, K.-H., Piezoelectric energy harvesters for biomedical applications. *Nano Energy*, 57, 879–902, 2019, https://doi.org/10.1016/j.nanoen.2019.01.012.

98. Elahi, H., Butt, Z., Eugnei, M., Gaudenzi, P., Israr, A., Effects of variable resistance on smart structures of cubic reconnaissance satellites in various thermal and frequency shocking conditions. *J. Mech. Sci. Technol.*, 31, 9, 4151–4157, 2017, https://doi.org/10.1007/s12206-017-0811-z.

99. Elahi, H., Eugeni, M., Gaudenzi, P., Qayyum, F., Swati, R.F., Khan, H.M., Response of piezoelectric materials on thermomechanical shocking and electrical shocking for aerospace applications. *Microsyst. Technol.*, 24, 9, 3791–3798, 2018, https://doi.org/10.1007/s00542-018-3856-8.

100. Rehman, W.U., Nawaz, H., Wang, S., Wang, X., Luo, Y., Yun, X., Iqbal, M.N., Zaheer, M.A., Azhar, I., Elahi, H., Trajectory based motion synchronization in a dissimilar redundant actuation system for a large civil aircraft, in: *2017 29th Chinese Control and Decision Conference (CCDC)*, IEEE, 2017, https://doi.org/10.1109/CCDC.2017.7979383.

101. Memmolo, V., Elahi, H., Eugeni, M., Monaco, E., Ricci, F., Pasquali, M., Gaudenzi, P., Experimental and numerical investigation of PZT response in composite structures with variable degradation levels. *J. Mater. Eng. Perform.*, 28, 6, 3239–3246, 2019, https://doi.org/10.1007/s11665-019-04011-4.

102. Rehman, W.U., Jiang, G., Wang, Y., Iqbal, N., Rehman, S.U., Bibi, S., Elahi, H., A new type of aerostatic thrust bearing controlled by high-speed pneumatic valve and a novel pressure transducer. *Int. J. Automot. Mech. Eng.*, 16, 4, 7430–7446, 2019, https://doi.org/10.15282/ijame.16.4.2019.16.0550.

103. Elahi, H., Eugeni, M., Gaudenzi, P., Design and performance evaluation of a piezoelectric aeroelastic energy harvester based on the limit cycle oscillation phenomenon. *Acta Astronaut.*, 157, 233–240, 2019, https://doi.org/10.1016/j.actaastro.2018.12.044.

104. Eugeni, M., Elahi, H., Fune, F., Lampani, L., Mastroddi, F., Romano, G.P., Gaudenzi, P., Experimental evaluation of piezoelectric energy harvester based on flag-flutter, in: *Conference of the Italian Association of Theoretical and Applied Mechanics*, Springer, 2019, https://doi.org/10.1007/978-3-030-41057-5_65.

105. Wang, J., Zhou, S., Zhang, Z., Yurchenko, D., High-performance piezoelectric wind energy harvester with Y-shaped attachments. *Energy Convers. Manage.*, 181, 645–652, 2019, https://doi.org/10.1016/j.enconman.2018.12.034.

106. Abdelkefi, A., Aeroelastic energy harvesting: A review. *Int. J. Eng. Sci.*, 100, 112–135, 2016, https://doi.org/10.1016/j.ijengsci.2015.10.006.

107. Kim, S., Jahandar, M., Jeong, J.H., Lim, D.C., Recent progress in solar cell technology for low-light indoor applications. *Curr. Altern. Energy*, 3, 1, 3–17, 2019, https://doi.org/10.2174/1570180816666190112141857.

108. Tsvetkov, N., Larina, L., Ku Kang, J., Shevaleevskiy, O., Sol-gel processed TiO2 nanotube photoelectrodes for dye-sensitized solar cells with enhanced photovoltaic performance. *Nanomaterials*, 10, 2, 296, 2020, https://doi.org/10.3390/nano10020296.

109. Elahi, H., Tamoor, A., Basit, A., Israr, A., Swati, R.F., Ahmad, S., Ghafoor, U., Shaban, M., Design and performance analysis of hybrid solar powered geyser in Islamabad, Pakistan. *Thermal Sci.*, 24, 2 Part A, 757–766, 2020, https://doi.org/10.2298/TSCI180311299E.

110. Cha, S., Kim, S.M., Kim, H., Ku, J., Sohn, J.I., Park, Y.J., Song, B.G., Jung, M.H., Lee, E.K., Choi, B.L., Porous PVDF as effective sonic wave driven nanogenerators. *Nano Lett.*, 11, 12, 5142–5147, 2011, https://doi.org/10.1021/nl202208n.

111. Yang, W., Chen, J., Zhu, G., Wen, X., Bai, P., Su, Y., Lin, Y., Wang, Z., Harvesting vibration energy by a triple-cantilever based triboelectric nanogenerator. *Nano Res.*, 6, 12, 880–886, 2013, https://doi.org/10.1007/s12274-013-0364-0.

112. Yuan, M., Cheng, L., Xu, Q., Wu, W., Bai, S., Gu, L., Wang, Z., Lu, J., Li, H., Qin, Y., Biocompatible nanogenerators through high piezoelectric coefficient 0.5 Ba (Zr0. 2Ti0. 8) O3-0.5 (Ba0. 7Ca0. 3) TiO3 nanowires for *in-vivo* applications. *Adv. Mater.*, 26, 44, 7432–7437, 2014, https://doi.org/10.1002/adma.201402868.

113. Mathuna, C.O., O'Donnell, T., Martinez-Catala, R.V., Rohan, J., O'Flynn, B., Energy scavenging for long-term deployable wireless sensor networks. *Talanta*, 75, 3, 613–623, 2008, https://doi.org/10.1016/j.talanta.2007.12.021.

114. Azevedo, J.A. and Santos, F., Energy harvesting from wind and water for autonomous wireless sensor nodes. *IET Circuits Devices Syst.*, 6, 6, 413–420, 2012, https://doi.org/10.1049/iet-cds.2011.0287.

115. Nasir, A.A., Zhou, X., Durrani, S., Kennedy, R.A., Relaying protocols for wireless energy harvesting and information processing. *IEEE Trans. Wireless Commun.*, 12, 7, 585–590, 2013, https://doi.org/10.1109/TWC.2013.062413.122042.

116. Sodano, H., Dereux, R., Simmers, G., Inman, D., Power harvesting using thermal gradients for recharging batteries, in: *Proceedings of 15th International Conference on Adaptive Structures and Technologies*, 2004.

11

Role of IoT in the Renewable Energy Sector

Veerakumar Chinnasamy* and Honghyun Cho

Department of Mechanical Engineering, Chosun University, Pilmundaero, Dong-gu, Gwangju, South Korea

Abstract

Implementing renewable energy systems is significantly increased world-wide to address various environmental issues. The capacity of renewable power source installation expanded in recent times and tends to increase in the future. Therefore, it is necessary to optimize renewable energy generation and system flexibility so that it can be utilized safely. Efficient utilization of renewable energy will result in minimized environmental impact and increased reliability on the power grid. Internet of Things (IoT) facilitates integrating the different renewable energy sources and helps in the uniform distribution of available energy to the targeted beneficiaries when needed. It also improves the energy efficiency of the renewable system by effective management of generation, transmission, supply, and demand. This chapter discusses and highlights the role of IoT in the renewable energy sector.

Keywords: Internet of Things, renewable energy, solar, smart grid, energy distribution

11.1   Introduction

The global energy demand is going up drastically, and the environmental impact of using fossil fuels is simultaneously increasing. Global warming due to CO_2 emissions from the energy sector alerted us to shift towards renewable energy utilization. There are several practical problems in completely utilizing the renewable energy source, and it is

*Corresponding author: veerakumar@chosun.ac.kr

R. Rajasekar, C. Moganapriya, P. Sathish Kumar and M. Harikrishna Kumar (eds.) Integration of Mechanical and Manufacturing Engineering with IoT: A Digital Transformation, (305–316) © 2023 Scrivener Publishing LLC

necessary to efficiently use it by high-level integration of monitoring, transmission, and distribution systems according to the generation and demand. Energy efficiency plays a key role in reducing energy consumption, which in turn reduces the emission from the energy sector. Energy efficiency can be achieved by effective management of data related to generation, transmission, and distribution along with the various designated consuming sectors, such as buildings, industries, and transportation.

11.2 Internet of Things (IoT)

IoT is a method of adapting advanced technologies with the aim of providing better connectivity between different devices. IoT uses different technologies to sense, record, and transmit real-time data for optimal decision-making. Based on these data, IoT can provide different services to the consumer efficiently. IoT has applications in different fields, such as transportation, healthcare, building energy management, etc. IoT can contribute significantly to the renewable energy sector by effective management of systems.

There are several steps in developing an IoT system for a specific application. The components of the IoT platform are shown in Figure 11.1. Initially, the devices for IoT systems are selected based on the required

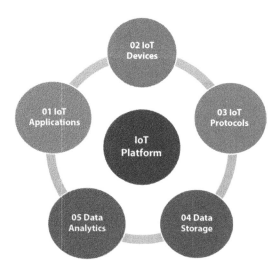

Figure 11.1 Components of IoT platform.

applications. Then the protocol for communication between the devices is selected according to the needs. The next step is to store the collected data from various devices according to the protocol. Finally, the stored data is processed in the form of real-time analytics.

11.3   IoT in the Renewable Energy Sector

About 80% of global final energy consumption is from fossil fuels. By following energy-efficient techniques and using renewable energy, the negative effect of using fossil fuels can be minimized. IoT helps to reduce the energy loss and controls the emission. The energy generation and consumption can be monitored in real time and improves performance. This section provides insight into IoT in various renewable energy sectors for its effective utilization.

11.3.1   Automation of Energy Generation

Automation in the power sector is necessary to monitor and control energy generation and distribution. The loss in generation and other various risks related to blackouts can be addressed through implementing IoT. Solar and wind are two major contributors among the different types of renewable energy sources. The reliability of these energy systems is increased by IoT systems. As a part of IoT, different sensors are being installed to automate the process and increase efficiency. For example, in solar photovoltaic power plants, a tracking system facilitates to harvest of maximum solar energy throughout the day. IoT systems control the tracking system by sensing solar radiation through sensors.

A remote IV tracing system of solar power facilities was studied by Shapsough et al. [1]. It is well known that solar energy is abundant in desert areas, and installing photovoltaic modules in such locations will produce a considerable amount of renewable energy. Accumulation of dust over the photovoltaic panels seriously affects the solar radiation incident on the photovoltaic panels. In this scenario, IV curves are used to evaluate the performance of photovoltaic panels. An IoT-based solar monitoring system is designed and implemented to monitor the loss due to soil accumulation. A solar monitoring architecture based on the IoT paradigm is shown in Figure 11.2.

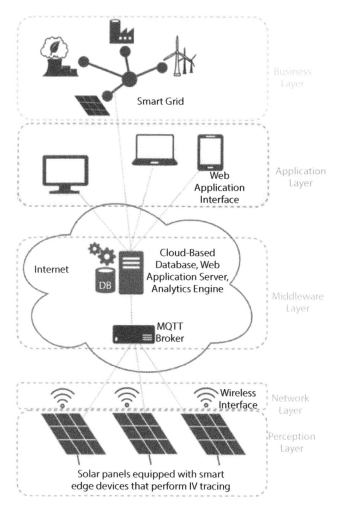

Figure 11.2 A solar monitoring architecture based on the IoT paradigm.

P. de Arquer Fernandez *et al.* [2] discussed a 3 MW photovoltaic power plant in Spain. More data are generated from these plants for monitoring the power plant, and it needs additional PLCs to handle the communication between the devices while adopting Figure 11.3 (a). When an IoT system is integrated, as in Figure 11.3 (b), independent communications are possible, and data read speed also increases.

A block diagram of IoT-based solar energy harvesting solution offered to a customer by Embitel Technologies (I) Pvt Ltd, Gujarat, India, is shown in Figure 11.4 [3].

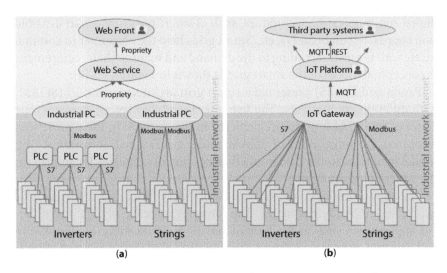

Figure 11.3 Monitoring system in the plant (a) before IoT integration, (b) after IoT integration.

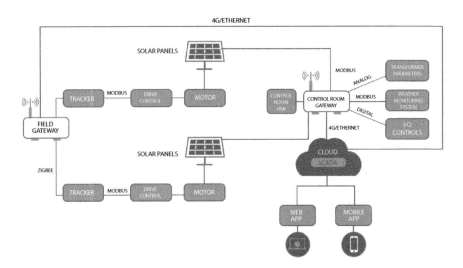

Figure 11.4 IoT-based solar energy harvesting solution.

11.3.2 Smart Grids

The major drawback in utilizing renewable energy is its lack of reliability in transmission and distribution. The energy grids which are used traditionally are capable of one-way power transmission. Smart grids are electrical transmission systems that enable two-way power transmission. It also consists of digital communication techniques, such as smart metering, smart distribution boards, circuit breakers, etc. Smart grids help the consumer to communicate with the grid according to the demand and manage their requirement. A conceptual model of the smart grid is shown in Figure 11.5 [4].

Panda and Das [5] presented a smart grid architecture model (SGAM) with different components. The five interoperability layers, domains, and zones of the SGAM model are depicted in Figure 11.6. In their work, they discussed the integration of SGAM under operational constraints. This model facilitates to the design of efficient control strategies for improving the system performance.

The application of conservative power theory (CPT) for the smart grid was reviewed by Ding *et al.* [6]. The future perspective of CPT and different applications such as load characteristic identification, accountability, and revenue, power converters are also discussed and presented. Rohde and Hielscher [7] discussed the smart grid development in Germany and analyzed the challenges in smart grid technology. They also developed a conceptual framework for tracking the smart grid field changes. Yapa *et al.* surveyed blockchains for future smart grids. Figure 11.7 illustrates the evolution of smart grids from conventional to smart grid 2.0.

The first-generation smart grid helps in bidirectional communication by advanced information technologies to remotely monitor the energy transfer. Smart grid 2.0 facilitates seamless connectivity of different types

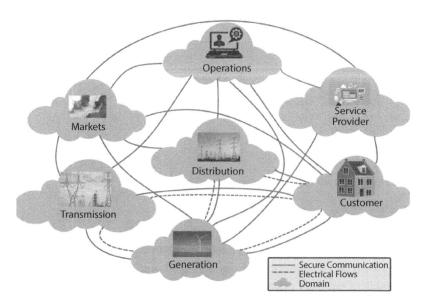

Figure 11.5 A conceptual model of smart grid.

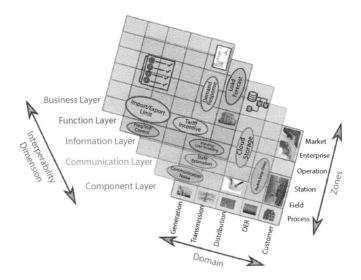

Figure 11.6 Five interoperability layers, domains, and zones of the SGAM model.

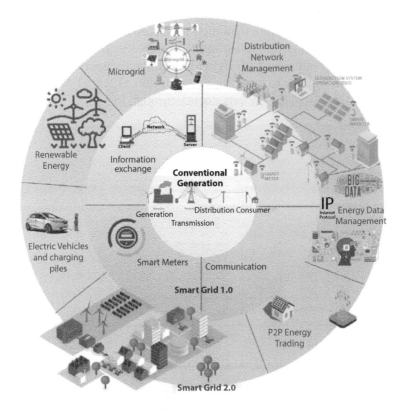

Figure 11.7 Evolution of smart grid 2.0.

of power generation that include renewable power systems. Renewable power generation systems, such as photovoltaic plants in remote locations, can be connected with the grid and load management centers.

11.3.3 IoT Increases the Renewable Energy Use

IoT and smart grids improved the renewable energy utilization capacity. The generation, transmission, and distribution of renewable power become easy, and it also monitors the power consumption. IoT facilitates controlling the power plant according to the requirement by receiving real-time data from the consumer.

11.3.4 Consumer Contribution

Nowadays, consumers started using renewable power sources to meet their energy needs in the aim of reducing their electricity expenses. The subsidies provided by the governments for renewable power generation attracts the people to show interest in renewable energy for their requirement. The excess power generated more than their needs can be supplied to the smart grid system in exchange for money and carbon credits. This helps to increase renewable energy utilization, which in turn helps to reduce the adverse effect on the environment due to fossil fuel consumption. Figure 11.8 shows smart homes with renewable energy dependence through IoT [8].

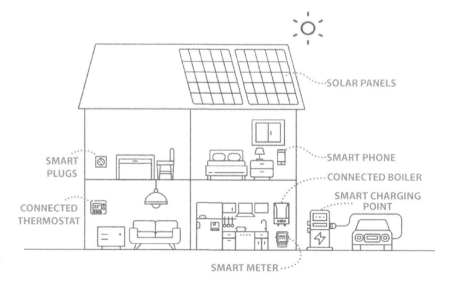

Figure 11.8 IoT and smart homes.

11.3.5   Balancing Supply and Demand

The supply and demand of renewable power can be monitored by using smart control systems. These control systems trigger the sensors according to the needs of the consumer, and the power supply will be increased or decreased based on the input feed from the devices. In order to achieve the demand side management, IoT operating with smart devices helps to ensure the requirement of power for the end user. The energy requirement is optimized based on the inputs from various systems connected to the load, and the demand is fulfilled by altering the generation and supply. The smart grids connected with smart devices from the demand and supply end are shown in Figure 11.9.

11.3.6   Smart Buildings

Buildings are one of the major energy-consuming structures throughout its operation. The quantity of consumption varies in different types of buildings, such as domestic and commercial buildings. The energy consumption in domestic buildings is mainly for heating, ventilation, and

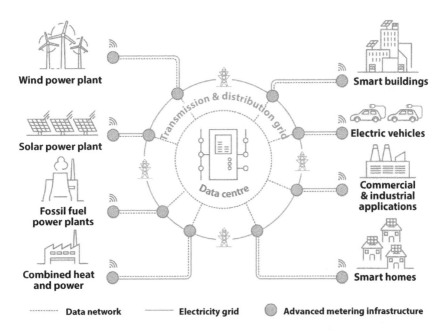

Figure 11.9 Smart grids connected with smart devices from the demand and supply end.

air-conditioning (HVAC) [9], cooking, lighting, and hot water supply. IoT helps to control the HVAC system to perform efficiently according to the load and reduce the electricity consumption. The lighting system is also integrated and controlled by an IoT system based on the requirement. It also shifts the electricity load by peak and off-peak control strategies. Intelligent devices that control energy-consuming devices can avoid losses in various parts of the buildings [10].

11.3.7   Smart Cities

Due to the increased living standards of people, the urbanization and development of smart cities are in practice to utilize the available energy effectively. Development in IoT technologies helps to address the problem raised in the smart city concept [11]. The residential houses, power generation units, and other industries are connected together through IoT, and the data about their energy utilization is monitored. This helps to reduce environmental degradation through water and air pollutions [12]. With the help of IoT technologies, smart city management can be successfully implemented by collecting information, monitoring, efficient generation, and transmission, etc. [13].

11.3.8   Cost-Effectiveness

The IoT integrated renewable energy systems play an important role in cost-cutting by efficient power management strategies. The peak and off-peak loads are categorized, and this data is transmitted to the renewable power generation unit. According to the received data, the power plants will generate the required power and supply to the grid. This helps to balance the excess load during the peak time and also reduces the power cost by supplying the power from a renewable source.

11.4   Data Analytics

11.4.1   Data Forecasting

Data forecasting is a major outcome of implementing IoT technologies that monitor and records the data from power generation to consumption along with transmission. These data will be transmitted to the control system. The power consumption can be altered based on the cost of power available at that instant. Forecasting helps to avoid problems related to demand, overload, and failure [14].

11.4.2   Safety and Reliability

The data analytics of the IoT helps to achieve better safety and reliability on the renewable power source. It helps to avoid power leakage and loss during transmission. The failures during natural or artificial disasters can easily be identified and rectified to provide an uninterrupted power supply [15]. The theft of the electrical parts in the transmission grid can be avoided as the IoT system monitors the devices connected to the grid, and if any, disconnection or unusual activity around the installed premise will be informed to the control center.

11.5   Conclusion

Renewable energy systems contribute to the power sector considerably, and its limit has been increasing every day. Advanced technologies, such as IoT, help to enhance the performance of the renewable system and increases its usage. It also facilitates the consumer to generate and utilize renewable power in an optimized way. In this chapter, the role of IoT in the renewable energy sector is discussed, and its contribution in different sections from generation to utilization is briefly presented. The application of IoT in different areas such as smart grids, smart cities, and buildings is also discussed. With the help of IoT, effective energy management is achieved that will help to reduce energy loss. Overall, it can be concluded that the development of IoT technologies is essential for the smart and efficient usage of renewable energy.

References

1. Shapsough, S., Takrouri, M., Dhaouadi, R., Zualkernan, I., An IoT-based remote IV tracing system for analysis of city-wide solar power facilities. *Sustain. Cities Soc.*, 57, 102041, 2020.
2. de Arquer Fernández, P., Fernández Fernández, M.Á., Carús Candás, J.L., Arboleya Arboleya, P., An IoT open source platform for photovoltaic plants supervision. *Int. J. Electr. Power Energy Syst.*, 125, 106540, 2021.
3. Embitel, https://www.embitel.com/iot-casestudies/development-of-control-room-gateway-for-solar-tracking-system.

4. Arnold, G., FitzPatrick, G., Wollman, D., Nelson, T., Boynton, P., Koepke, G., Hefner Jr., A., Nguyen, C., Mazer, J., Prochaska, D., Swanson, M., Brewer, T., Pillitteri, V., Su, D., Golmie, N., Simmon, E., Eustis, A., Holmberg, D., Bushby, S., Janezic, M., Jillavenkatesa, A., NIST Framework and Roadmap for Smart Grid Interoperability Standards, Release 2.0, Special Publication (NIST SP), National Institute of Standards and Technology, Gaithersburg, MD, 2012.

5. Panda, D.K. and Das, S., Smart grid architecture model for control, optimization and data analytics of future power networks with more renewable energy. *J. Cleaner Prod.*, 301, 126877, 2021.

6. Ding, Y., Mao, M., Chang, L., Conservative power theory and its applications in modern smart grid: Review and prospect. *Appl. Energy*, 303, 117617, 2021.

7. Rohde, F. and Hielscher, S., Smart grids and institutional change: Emerging contestations between organisations over smart energy transitions. *Energy Res. Soc. Sci.*, 74, 101974, 2021.

8. Ηυανικων, Υ.Η.Λ.Μ., Ηυανικων, Κ.Α.Ι.Μ., Σων, Τ.Π., Σημασων, Σ.Ο.Τ., Ληροφορια, Μ.Δ.Σ.Α.Ο.Η.Π., Ια, Ι.Δ., and Υαραλαμπο, Σ.Ι.Θ., Internet of Things-Innovation landscape, in: *Cyber Resil. Syst. Networks*, (July 2016), pp. 1–150, 2019.

9. Vakiloroaya, V., Samali, B., Fakhar, A., Pishghadam, K., A review of different strategies for HVAC energy saving. *Energy Convers. Manage.*, 77, 738–754, 2014.

10. Arasteh, H., Hosseinnezhad, V., Loia, V., Tommasetti, A., Troisi, O., Shafie-Khah, M., Siano, P., IoT-based smart cities: A survey. *EEEIC 2016 - Int. Conf. Environ. Electr. Eng.*, pp. 2–7, 2016.

11. Ejaz, W., Naeem, M., Shahid, A., Anpalagan, A., Jo, M., Efficient energy management for the Internet of Things in smart cities. *IEEE Commun. Mag.*, 45, 1, 153–155, 2007.

12. Lagerspetz, E., Motlagh, N.H., Arbayani Zaidan, M., Fung, P.L., Mineraud, J., Varjonen, S., Siekkinen, M., Nurmi, P., Matsumi, Y., Tarkoma, S., Hussein, T., MegaSense: Feasibility of low-cost sensors for pollution hot-spot detection. *IEEE Int. Conf. Ind. Informatics*, vol. 2019-July, pp. 1083–1090, 2019.

13. Mohanty, S.P., Choppali, U., Kougianos, E., Everything you wanted to know about smart cities. *IEEE Consum. Electron. Mag.*, 5, 3, 60–70, 2016.

14. Vignesh, R. and Samydurai, A., A survey on IoT system for monitoring solar panel. *Int. J. Sci. Dev. Res.*, 1, 11, 114–115, 2016.

15. Gurav, U. and Patil, C., IoT based interactive controlling and monitoring system for home automation. *Int. J. Adv. Res. Comput. Eng. Technol.*, 5, 9, 2392–2396, 2016.

Index

Printed and bound by CPI Group (UK) Ltd, Croydon, CR0 4YY

27/10/2024

14580175-0003